Yale Agrarian Studies Series
JAMES C. SCOTT, *Series Editor*

"The Agrarian Studies Series at Yale University Press seeks to publish outstanding and original interdisciplinary work on agriculture and rural society—for any period, in any location. Works of daring that question existing paradigms and fill abstract categories with the lived-experience of rural people are especially encouraged."
JAMES C. SCOTT, *Series Editor*

For more information on the Yale Agrarian Studies Series, visit www.yalebooks.com.

The Nature of Entrustment

Intimacy, Exchange, and the
Sacred in Africa

Parker Shipton

Yale University Press   NEW HAVEN & LONDON

Designed by James J. Johnson and set in Ehrhardt Roman by Tseng Information Systems, Inc.
Printed in the United States of America by Sheridan Books.

Library of Congress Cataloging-in-Publication Data

Shipton, Parker MacDonald.
The nature of entrustment : intimacy, exchange, and the sacred in Africa / Parker Shipton.
p. cm. — (Yale agrarian studies series)
Includes bibliographical references and index.
ISBN: 978-0-300-11601-4 (cloth : alk. paper)
 1. Economic anthropology—Africa—Great Lakes Region. 2. Ceremonial exchange—Africa—Great
Lakes Region. 3. Luo (Kenyan and Tanzanian people)—Finance. 4. Luo (Kenyan and Tanzanian
people)—Commerce. 5. Luo (Kenyan and Tanzanian people)—Economic conditions. 6. Great Lakes
Region (Africa)—Economic conditions. 7. Great Lakes Region (African)—Social life and customs.
I. Title
GN645.S515 2007
306.3096—dc22

                                                                                2006033475

A catalogue record for this book is available from the British Library.

The paper in this book meets the guidelines for permanence and durability of the Committee
on Production Guidelines for Book Longevity of the Council on Library Resources.

10 9 8 7 6 5 4 3 2 1

TO
James Morris Shipton
AND TO THE MEMORY OF
Mary Elizabeth Cornu Shipton

Contents

Preface

Humans borrow and humans lend. We give, take, and give back all sorts of things: animal, vegetable, and mineral; large and small. A book to read, a sled to slide, a cart to move a sack of corn. We make friends with new neighbors by "borrowing" a cup of sugar, even one that all know will not be returned. We entrust an ox to a neighboring farmer, a house to a sitter, on the tacit or explicit understanding that at some point we will want them back. We devise commercial strategies like the letter of credit, and financial ones like the bond and the mortgage, that all boil down in one way or another to borrowing and lending. Surely we borrow and lend more things than any other species. It seems to be part of what makes us human.

We do it between kith and kin, as well as between strangers. Goethe put it this way: "Live in as small a circle as we will, we are all creditors or debtors before we have had time to look around" (*Elective Affinities* [Bk. II, ch. 4]). No one can go long without owing or being owed.

And yet sometimes we borrow or lend more than we want or later wish we had. Shakespeare warned us that borrowing and lending can jeopardize friendship and livelihood: "Neither a borrower nor a lender be; / For loan oft loses both itself and friend, / And borrowing dulls the edge of husbandry" (*Hamlet*, Act I, scene iii, 75–77). If Shakespeare and Goethe are *both* right, then managing our obligations to each other is a delicate matter indeed.

What can make it harder still is that not all the things that are bestowed upon us, entrusted to us, are meant to be given back. Sometimes our loans shade into other forms of exchange like gifts and bequests, in which it is not so clear whether our obligation is to give right back (like a borrowed axe), to tend well and wisely and maybe end up keeping (like a pet boarded indefinitely, or a child fostered), or to pass along as an heirloom (like an engagement ring or a soldier's medal). Arrangements like these are common—one might say normal and natural. But when they go wrong, as they so often seem to do, emotions run high, and these emotions are just as natural.

Who owes what to whom, and why? Can every credit be calculated, and need every debt, *could* every debt, ever be discharged? Is it best to settle up with our kith and kin before paying off unknown creditors—or to see to our duties to the living before those to the dead? Questions like these have few easy answers, and for centuries, scholars in Shakespeare's and Goethe's traditions have been poring over their texts looking for clues. What, they ask, does a hapless Antonio really owe a heartless Shylock if barriers of category and convention divide them? Why can't there be more Portias to warm cold hearts? What does a Hamlet owe his father once the father is dead?

Unfortunately, these authors and their characters—some of whom they wrap in ethnic, religious, or gender stereotypes—can guide us only so far. Neither writer tells us in his aphorism whether borrowing can bind strangers, whether debts must connect continents, or how to translate "obligation" between tongues. Neither really tells us whether those we owe, and those who owe us, include the dead and unborn as well as the living.

The questions are not just literary ones but cultural and moral too, and in our times they are pressing. Lending over long distances, it is easy to say, weaves webs of obligation between places that hitherto had few common ties. But that does not mean that everyone, everywhere, understands borrowing and lending in the same way.

Who, then, feels beholden to whom? Anyone interested in culture in our times has a chance, perhaps a duty, to ask such questions about parts of the world that for Shakespeare or Goethe were left blank on maps or illustrated with fanciful monsters. If one cannot find out what we owe our kith or kin, let alone the distant, the dead, or the divine—then at least one might hope to learn something about how other people in different regions approach the basic problem of human indebtedness. One might learn, for instance, whether *they*, when in small circles, become creditors or debtors before they have had time to look around; what they think constitutes an obligation; how they manage to cope with credits and debts that tie them in webs; and whether those webs include an "us."

The pages that follow describe one of those places distant from where

I now write, to show what people there might teach about whether, how, and how cheerfully to be borrowers and lenders. We may expect, as with any such adventure, to have to recast before we are done the ideas with which we began — maybe including how we define "us" and "them."

This study is one of three in roughly simultaneous preparation, an informal trilogy written to be readable as single volumes and comprehensible in any sequence. In all of them, the case of the Luo people and Kenyan nation where most of them live play a central role; and in all of them credit and debt, among other forms of entrustment, are at issue. But the discussions and implications reach well beyond Africa and that topic.

The aims for this first part of the triad are simple. At the most basic, I hope here to show that credit and debt are cultural as much as economic or political. As the topic involves trust and entrustment, I call it fiduciary culture: shared, learned ways of thinking and acting that involve some sense of obligation. Some of it takes reciprocal forms, but some involves serial transfers passed between generations. I show how some objects, and animals and persons too, move around in a rural African setting, and suggest why; but these pages are really more about the people than the things themselves. I try to illustrate some kinds of borrowing, lending, and passing along that occur in rural Africa, and between town and country, without anyone's direct help from outside, and often without money. Some of these forms of entrustment make up part of reproductive, ritual, and symbolic life, and people feel strongly about them. These are dealings and sentiments of which foreign financiers and aid planners are often unaware when they arrive with loan schemes for people they assume too poor and ignorant to have had access to credit before. To describe these local processes means introducing a few different ways of thinking about borrowing and lending, about entrustment and obligation, and looking more carefully at how we frame such ideas in order to communicate. Touching on several domains of experience, these pages show how hard it is to distinguish activities that are economic from ones that are not. This ethnographic background sets the cultural stage for the two companion volumes on land tenure and on financial credit.

The second volume in the works, focusing on land, burial, and the mortgage, treats the topic of property and what I call ideologies of attachment, describing some long-term and recurring misunderstanding from a mixing of cultures in rural African land affairs. It first traces some roots of a Euro-American understanding of capital, and a theory of property. It then discusses the history and nature of land settlement in the Luo country, and the system of landholding that has evolved there. It examines the state-sponsored imposition of private property in land over the last half century, and the use of land as collateral for loans. Describing the working and outcome of a freehold-mortgage system of land tenure and finance in the Luo country and Kenya, I question whether national and foreign developers have made inappropriate assumptions about land and loyalty, and about the relation between time and money, in these East African settings and in other parts of the world.

A third volume, about credit between cultures, reviews other national and international attempts at rural finance — without land mortgages — and ways Luo and others nearby have participated in and responded to them. Public, private, and self-help approaches are described and compared. A singularly ambitious aid program, sponsored by the World Bank and other agencies, is analyzed in detail to show how Luo people can transmute and manipulate initiatives from outside their country, putting these to their own purposes. In this third work especially, I attempt to show how experiences repeat, and why. The book also looks beyond aid and development to show other options rural-dwelling people in Africa exercise to solve poverty and related problems and to transcend what they cannot solve. It suggests what others can do to honor, accommodate, or at least comprehend those African aspirations.

Together, these three sister studies attempt to provide a multi-angled view of borrowing, lending, and passing along. They also combine in supporting an argument about the nature of international interventions over the past half century or so, and about what needs to be reconceived and done differently in the future to make the outcomes of long-distance relations happier for African people concerned. Matters that seem on the surface to be economic and technical turn out to be profoundly cultural and symbolic too. Issues that seem to be about ephemeral configurations of bureaucratic and market actions turn out to be about timeless truths and sacred cycles as well. To some, the stakes are nothing less than life itself; and their means of redress, when pressed, include violence. To most, the outcomes are slower and more subtle, though no less dramatic. To understand the processes at work, and the effects of some intercultural dealings on rural lives, we shall need to take into account long-distance connections and the passage of time over generations, and to respect African ingenuity, wisdom, and dignity.

Sometimes ends become beginnings, and beginnings ends. This is the first of the three related studies to be published, but it was the last to be drafted. When I began studying land tenure and credit in East Africa, I had little idea that child fostering, marriage and funeral transfers, or ritual sacrifice would eventually come to the center of my screen. (My manual typewriter had no screen.) A question like what we ultimately owe to whom, or why, would then have seemed too imponderable to be worth asking. Nor had I yet learned—African people had not yet taught me—the importance of intergenerational legacies or stewardship, or the importance of finding our place between those who have gone before and those who will come after, as though both were, or are, right close at hand.

What is described in these pages are subjects whose meanings and specifics I did not always fully appreciate at first in situ but was taught about as I went. Some had to be filled in over many years' reading, repeat visits to East Africa and other parts of the continent, and sustained contact with savvy African sojourners abroad. They are more basic, more enduring, less amenable to budgets or schedules than finance or land titling programs. They also set a broad backdrop of an African culture and society for those unfamiliar. That is why I have put them first. But beginnings and ends sometimes have a way of connecting, and it seems to me now that the kinds of topics in this book, as elusive or imponderable as they might seem at first, have more than a little to do with those other, more ostensibly "practical" topics as well, that so preoccupy outsiders wanting to do something about conditions in Africa. How and why they do will take some explaining.

The three studies together—a trilogy with at small "t"—allow a critical perspective on an international reflex of lending, and on an imposed property regime. In particular, they cast doubt on the institution of the mortgage. This system, combining loan, clock, and private land title, is a rather recent innovation in many parts of Africa and the tropics where tried or planned. By worsening social division and landlessness, it threatens further damage to lives lately under stress in agrarian settings, and to continue doing so insidiously over generations. But that need not happen if one begins by asking what is there and working already, instead of what is lacking and ought to be imposed or reformed. If this first brief volume helps color in a part of the map, illustrating the role of fiduciary culture in an African way of life, it will have achieved its main aim. If it also reminds a reader how much we all owe each other, and how futile is the notion that any of us, as an individual, will ever be economically, politically, or morally self-sufficient, so much the better.

Acknowledgments

Some debts can never be directly repaid, but it is an honor for me to acknowledge the many who have contributed their care, help, and expertise to this work — and a regret that space limitation prevents my naming them all here.

Generous financial support has come from several sources. A British Marshall Scholarship made possible the preliminary research at Oxford. Grants for further research at Cambridge, and in Kenya between late 1980 and early 1983, were provided by St. John's College, Cambridge, and its Warmington Research Studentship, the Smuts Memorial Fund of Cambridge University, the Wenner-Gren Foundation for Anthropological Research, and the Radcliffe-Brown Memorial Fund of the Royal Anthropological Institute of Great Britain and Ireland. The Social Science Research Council and the American Council of Learned Societies provided a fellowship for a revisit to Kenya and for library research. Two research and writing fellowships afforded precious time and inspiring colleagueship: those of the Carter G. Woodson Institute for Afro-American and African Studies of the University of Virginia (with added travel funds for archival research) in 1989–90, and of the Yale University Program in Agrarian Studies in 1992.

I am grateful to the University of Nairobi's Institute of African Studies and its Institute of Development Studies, which both warmly welcomed me as a research associate from 1980 to 1983 and in 1991 to 1992 respectively. For research clearances in Kenya I thank the Office of the President and the local administration; the Ministry of Agriculture, and the Ministry of Cooperative Development. Many officers of the World Bank, the U.S. Agency for International Development, the Kenyan Ministries of Agriculture and Cooperative Development, the Canadian International Development Agency, the Church Missionary Society, the British Council, the British Institute in East Africa, the Nairobi YMCA, the Rapogi Catholic Mission, and the National Museums of Kenya have generously lent their help in Kenya.

My colleagues and students at Harvard from 1984 on, and at Boston University, my main base since 1994, have lent me more support in the departments, centers, libraries, and classrooms than I can properly acknowledge here. The departments of anthropology at both universities, the former Harvard Institute for International Development, Harvard's John Winthrop House, and the Boston University African Studies Center have provided colleagues both stimulating and congenial. The Colleges of Arts and Sciences at both institutions have allowed me leaves for research and writing.

For kind permission to reprint revised parts of my article "Luo Entrustment," originally published in the journal *Africa*, I am grateful to the International African Institute and Edinburgh University Press.

Now some more personal debts. For more than two decades, and on three continents, Polly Steele has enriched this work with her companionship and her insightful contributions intellectual, moral, and marital. Sharing adventures, risking her health at times, and lending patience and persuasion, she has done more—for this work and for me—than words can tell.

In western Kenya, people of Kanyamkago, and of Kagan, Sakwa, Uyoma, and Isukha have shared their endless cheer and patience as my hosts, guides, friends, and teachers. Special thanks are due to Janes Athiambo Omolo, Joseph Obura Opon, and Emin Ochieng' Opere, who have all infused themselves into this work. They and for shorter periods Dan Odhiambo Opon, Meshack Magack Opon, Everlyn Millicent Adhiambo Odhiambo, Ben Omondi, Elly Owino Nyanje, and Fred Shijenje have all supplied their companionship, ingenuity, hard work, and vital contacts in their home neighborhoods. They and kin, including Rispa Akech, Esther Akinyi, Charles Opon, and Ulda Ayier in the elder generation, touchingly bent their genealogies to fit me in. Ismael Owiro, Joash Songa, and Hudson Azangu thoughtfully arranged housing when I was not living with families. Jeremiah and Jedidah Okumu offered kind hospitality in Kagan, as did Agnes Azangu Tizika and family in Isukha. The staff and students of the school I have called Kagogo Primary School and their families proved lively, supportive neighbors and friends. Countless others offered shelter, hens, meals, jokes, and blessings.

These people have made my stays in and around Nyanza uniquely enjoyable and instructive. May this work be of interest to their children. *Ero kamano* and *asanteni.*

Frank Fitch and Elijah Mugo helped with transport, and Joseph Fitzsimons eased my first entry into Nyanza. John and Jennie Coope, Mark Harris, and David Salter provided entrée to many things in Nairobi, Valerie Smith in Maragoli Luhya country, and Richard Waller in Maasai country.

Debts to teachers and mentors seem only to grow with time. At Oxford Wendy James, Peter Rivière, Godfrey and Peter Lienhardt, Rodney Needham, and especially John Beattie taught me much about Africa and anthropology. So too, at Cambridge, did Jack Goody, Keith Hart, Polly Hill, John Iliffe, John Lonsdale, Sandy (A. F.) Robertson, Malcolm Ruel, Michel Verdon, and especially Ray (R. G.) Abrahams. The cohorts of students they taught, my co-initiates into a discipline they taught us to go beyond, were remarkable company. Elsewhere Rada Dyson-Hudson, Davydd Greenwood, Thomas Gregor, Richard Harfst, Bernd Lambert, and Migwe Thuo merit special mention.

One of the constant pleasures of working in and on western Kenya has been the superb colleagueship. It is the preferred East African way to acknowledge elders first, and I hope my peers and juniors will accept my letting this round of thanks go mainly to those who arrived ahead and cleared ground. Margaret Jean Hay, Robert and Sarah LeVine, John Lonsdale, John Middleton, Shem Migot-Adholla, Asenath Bole Odaga, H. W. O. Okoth-Ogendo, E. S. Atieno Odhiambo, David Parkin, Malcolm Ruel, Walter Sangree, and Aidan Southall deserve special thanks for so warmly welcoming me in and nurturing this work along. Of the scholarly contemporaries hoeing their own rows in Kenya alongside mine, I am particularly grateful to Patrick Alila, David Anderson, Maria Cattell, Michael Dietler, Angelique Haugerud, Ingrid Herbich, Elizabeth Munday Otunnu, Nancy Schwartz, Suzanne Mueller, and Richard Waller, who have variously advised me on clearances, shared wheels or daydreams, sent streams of clippings, commented on drafts, or otherwise gone far out of their way on my behalf.

At Harvard and at Boston University David Cole, Allan Hoben, David Maybury-Lewis, Sally Falk Moore, and Marguerite Robinson all encouraged me on from a generation ahead while they were buried under my drafts. Thomas Barfield, Charles Lindholm, Jane Guyer, and Pauline Peters each helped on both sides of the river; James McCann, James Pritchett, Diana Wylie, and Jean Hay (whose scissor pruning added much) shared daily ups and downs during writing and revision; and so many of my other fellow teachers also enlivened the corridors and seminars. Gretchen Walsh and Gene De-Vita spirited obscure sources out of their libraries while others who mined the stacks on my behalf included Michael Katz, Fred Klaits, Anne Lewinson, and Michael West. Maddalena Goodwin, Joanne Hart, Kathy Kwasnica, Janet O'Neil, and at Yale Kay Mansfield lent their managerial gifts to shepherd me through bureaucratic mazes and make my time at their universities enjoyable. Karyn Nelson produced the map and Michael Hamilton processed the photos; and Hormoz Goodarzy, Larry Andrus, and associated computer wizards rescued chapters from cyberdeath. Throughout, my students have taught me well.

In periods when I could not be in western Kenya Benjamin Campbell, Elizabeth Francis, Cynthia Hoehler-Fatton, Les Kaufman, Benzburg Nyakwana, Justus Ogembo, Herine Ogutha, Lisasa Opuka, Daivi Rodima, Elizabeth Siwo, and others brought it to Boston as if by the netful. Antonia Lovelace, Clancy Pegg, and Kathleen Hanson worked hard to put the quantifiable parts of survey interview data into usable form for this and other related work. Margaret Nipson, Suzanne Sloan, Katherine Yost, and others to be named separately all turned office tasking not just productive but convivial. Peter Agree, Fredrik Barth, and Jane Huber have lent friendly and wise advice on publishers.

Series editor James Scott and in-house senior editor Jean Thomson Black at Yale University Press have kindly and gamely supported my work, and I thank Jean and also Laura Davulis, Ann-Marie Imbornoni, and manuscript editor Joyce Ippolito for the skill they have shown in guiding this volume through the process. Diana Witt composed the index, and Debra Fried did the proofreading. The manuscript's anonymous reviewers provided comments and suggestions that have also improved it much.

James Shipton and Mary Elizabeth Cornu Shipton provided constant encouragement, suggestions, and proverbs when most parents would have deemed their job long done. Susannah has added joy.

To all these people, and to those I have neglected to mention, my lasting thanks. While it is customary, and true enough, to say that none of these shares the blame for any shortcomings, all share credit for whatever is worthy in this work.

CHAPTER 1

Introduction

One person throws a ball; the other does not know: whether he is supposed to
throw it back, or throw it to a third person, or leave it on the ground, or pick it
up and put it in his pocket . . .
—LUDWIG WITTGENSTEIN

Birthplace of humanity it may be. A home of the lion and leopard, the hippo, and the crowned crane it has been for sure. For
me though, the area had another appeal when I first decided to venture to
the equator to live among people overlooking Africa's largest lake. There was
something mysterious about the map.

Here was a straight border dividing an ostensibly capitalist Kenya from
an ostensibly socialist Tanzania. Or so they were proclaimed in the popular
press, and in government policies endlessly repeated in presidential speeches
and national development plans. Stretching southeast from the lake, the line,
first etched by European imperial statesmen in Berlin in 1884–85, cut right
across the territories of several population groups, whether labeled as races,
tribes, ethnic groups, or languages—as if these could all mean the same thing.

The people on the map nearest the lake, as you looked southeast along the
border, were known and still are as Luo, Kuria, and Maasai. The straight line
officially cut each of these unofficial territories in two. It was as if the people
there, in each of the three "tribes" (or non-tribes), were somehow of divided
minds about capitalism and socialism, those imported ideologies of which
Kenya and Tanzania had been standing since the early 1960s as Africa's famous exemplars. Were "the Luo" (or "the Kuria," or "the Maasai") on one
side of the border all busy entrepreneurs, buying and selling commodities and
each other in the chancy pursuit of selfish accumulation? Were their cousins
on the other side all comrades locked arm in arm, struggling to level conflicting classes, and freely sharing all they had? Or was this entire region of

eastern middle Africa a land of illegality and civil disobedience? There must be something the map was not showing. Something the speeches were not telling.

Odd too were the descriptions of the people written by earlier visitors, who seemed deeply divided by disciplines. The reports by the anthropologists about clan segmentation, ghostly vengeance, and initiation rites; by economists about crop price elasticity and interest rate spreads; and by political scientists about party platforms and nation building might as well have been describing three different continents, not to say planets. So here was a land divided by nation, by ethnicity, and in the journals by discipline.

The politicians' speeches and economists' five-year plans were all about growth and development; and for a while, after the independence of these countries, the hopes had seemed grounded in opportunities newly opened to African minds and hands. But by the end of the 1970s, talk of development was starting to ring hollow. Charts on trade and investment—or health and nutrition—were sloping downward. Nor did much of the rhetoric square with the stories and pictures in the papers. East African poverty seemed to rub right up against great wealth—at least in the towns or at the edges of the reserves or resettlement areas—without their canceling each other out. Production, entitlement, or charity was falling short.

The topic that drew me deeper into the library was attachment to land: a question of belonging. Everyone knew there were people in East Africa who did not quite belong, or who at least sometimes behaved as if they did not. There on the old colonial map of western and central Kenya were the so-called white highlands, vast tracts thinly settled by British and other immigrant settlers after the people who had lived there before had been lured, or forced, away from their homes, some for good. Then, as everyone knew, there were game reserves like the Tanzanian Serengeti and its Kenyan extension, the Maasai Mara, where Maasai people and their cows, goats, and sheep had been invited out—not always to stay, though—and tourists and their money invited in. But what about everyone else in the rest of the lands, labeled "native reserves"—for instance, around the lake and along that straight border? On what basis did *they* belong where they lived?

Colonial ethnographies revealed more than one system of land settlement, tenure, and community organization described as traditional. Around the eastern side of the lake, the Luo and their neighbors in western Kenya had a system in which kin groups, especially clans and lineages, seemed the most salient traditional principles of organization.[1] Around the southern side of the lake where Sukuma people and neighboring groups lived, by contrast, it was territorially defined chiefdoms and sub-chiefdoms that seemed the salient mode. Over the lake's horizon north and northwest, in Buganda and beyond in Bunyoro, had been those famous grand kingdoms, and systems of labor and loyalty sometimes described as "feudal." Curiously, the places with centralized kingdoms were not the ones with the highest population densities. The densest rural areas, such as those of the Luo and their Bantu-speaking Luhya and Gusii neighbors, were ones that had nothing of the sort. On what basis they *were* organized was still hard to tell. Whether centralized and uncentralized, or stratified and egalitarian, meant civilization and savagery I had learned to doubt, as everyone should. But it was already clear that some areas had long done without structures of rule that held others together. It was clear too that whatever sorts of community the European and postcolonial "capitalist" and "socialist" systems were imposed upon in the lake region, they took more than one form.

So there was more than one old way people tried to organize themselves, and more than one new. But how did old *connect* to new, I wondered—and how did rich connect to poor? What had European conquest, colonialism, and independence actually changed on the ground? What had become of "custom," and what was the nature of community? There had to be a way to find out such things by studying something specific. Something that related to the wealth and poverty, and the capitalism and socialism, that seemed so central to Africa's concerns, or to the world's concerns about Africa. Exchange, for instance. Gift. Theft. Whatever comes between.

But what *does* come between? Is there any form of transfer or exchange that can span the range from giving to taking, from sharing to profiteering?

If there is, I supposed, it must be the loan. Some sort of reciprocity. Something like a ball tossed back and forth—but one that might change size, shape, or feel between the giving and the taking back. A loan, or credit, can be made to help *or* to exploit—or maybe to do both at once. Or just to establish a relationship. And if any sort of exchange can bridge the gap between the most intimate relations within households even, and the most distant between continents, might it not again be the loan?

Eventually, people in the lake basin would teach me to ask whether credit relations, fiduciary relations, could do more than that: whether they could mean transferring living and breathing creatures, including humans themselves, whether they could bridge the living and the dead, or whether they could link the sacred and secular. But asking questions like these would require expanding "loan" or "credit" to a looser, more inclusive idiom, one like entrustment and obligation, in which reciprocity might not be the only way to understand things.

Attachment and loyalty would certainly be key concerns. To whom, and what, do people feel they **belong?** What might their transactions suggest about whom they identify with and trust . . . and what they owe to whom? And how might loyalty, trust, and obligation affect the nature of overseas aid (or "aid") interventions—something that must concern us all? If international aid agencies and national governments were attempting to change the basic nature of attachment and belonging through land tenure "reforms" (as they were and are), and to alter the nature of commodity flows through gifts and loans (as they also were and are), then how might their interventions relate to the cultural, social, and economic life of the region's people? These questions would take years, even decades, to reframe and address.

But now there was at least a topic: credit, debt, and human belonging. Belonging would mean attachment not just between people and things but also between people and each other.

On the Ground

As it turned out, when I arrived in Kenya in 1980 for my extended stay and my research there was no legal access across the border between Kenya and Tanzania. That road block turned out to be a blessing in disguise. Proceeding with the simple assumption that Kenya was about capitalism and Tanzania about socialism, and carrying out a neat comparison as such, would have meant missing most of what I was eventually to learn about how African rural economies really work.

Borderlines or not, people in the Luo part of the lake basin, where I settled, seemed to be living life more or less intact. Most were not behaving as capitalists *or* socialists in any consistent manner but were forging their own way, as indeed they had always done. They may have been members of a state, and by law a picture of the president hung high in every mud and wattle shop, to remind them—but that was a president whose regime's legitimacy would come increasingly into question, and who before two years would disappear for several days in an attempted coup. They were eating and using resources they grew or caught, and others they borrowed or bought. People who had been to school were deftly switching languages mid-sentence in the marketplace, falling back on simplified up-country Swahili or on English when their own languages failed. People were sharing customs of clay pot decorating or calabash mending, I would come to learn, and consulting healers from around the countryside or overseas to work their most powerful medicine among them as strangers. These things could be seen in my initial weeks in Nairobi and several more in the small town of Awendo, where there was a sugar plantation and factory and a strip of shops, some immigrant-owned: a busy, dusty sort of town with loudly competing music boxes, striving to outgrow its smallness. By then I was introduced to, and ready for, a more settled local community in the countryside.

My main base for the two years or so I would spend in western Kenya, and for the longest of my return visits a decade later, was a farming hamlet (which I shall call Kagogo) in Kanyamkago location, in that space between the main road and the great lake where there was little on the map. On the ground it was hill and valley land, mostly green, with scattered homesteads, and with farmed fields checkered around the landscape and visible from most anywhere.[2] It was just big enough to have a small school and a couple of one-window, mud-walled shops open on and off. Most of the houses on the hillside had thatched roofs and floors of bare earth; a few had corrugated iron roofs or cement floors. Some of the farmers' fields had sugar cane being grown for a factory at Awendo or for making molasses (jaggery), and some had tobacco they were growing for a multinational firm with a buying station nearby. But mostly it was maize, sorghum, and other annual crops (sometimes eaten in seed by birds in the fields) . . . and cattle (no longer raided by lions), sheep, goats, and chickens (still occasionally prey to smaller wildcats or hyenas). ·

From that main base, over the next two years or so, I made briefer, interspersed stays, up to several weeks at a time, in other rural settings. These included the locations of Kagan (where I stayed in a small market town) and lakeside Uyoma (where I stayed in a rural school), farther north in the Luo part of Nyanza Province. An eventual opportunity to sojourn among Luhya people in Isukha, north of the Luo country, gave me a chance to go visit a higher-altitude zone. These, and short visits to sites of other researchers and aid volunteers in Kuria and Maasai country and elsewhere in Kenya (as some of them visited mine too), and periodic short trips into the (then) South Nyanza district capital Homa Bay, the provincial capital Kisumu City, the national capital Nairobi, and other cities rounded out my view of Kenya and especially its agrarian western lakeside region.

This, then, was what the map looked like on the ground—and where I arrived to make friends (the crucial thing on which so much else depended), to learn a language and work along on another, to conduct interviews and observations daily, and to try to fit in as best I could. With a good deal of locally recruited help, while carrying on my own interactions, I eventually designed, participated in, and oversaw interview surveys of a more formal sort too, both around Kagogo in Kanyamkago and elsewhere in and nearby the Luo country, with numbers up to 286 informants representing their respective homesteads in the Kagogo case. More is said on these method topics toward the end of Chapter 3 and in the accompanying studies to come.

It did not take long to see I would be well treated in Kenya. From the start, wherever I stayed, my hosts, friends, and eventually fictive kin would be friendly, patient, and agreeable almost always. Indeed, where I stayed longest, they often enough proved warm and caring in all I could tell of their feelings toward me, and the same toward my spouse, who, when I had been there a year, came to Africa to join me and to work as a teacher and editor in Nairobi. They showed us this spirit in those years of my first visits (about two years and three months, covering all seasons, from October 1980 to January 1983, interrupted for a few weeks by illness) and in shorter revisits in 1986, 1988, and from October 1991 to January 1992.

Equatorial Africa is a land with much humor and good cheer at all but times of hunger or civil strife, and sometimes even then, as I would find out. Often people of little linguistic commonality laughed together — in that universal language, inflected so many ways to mean so many things. Sometimes they could not help laughing at the naïve questions a researcher in training asked — to which the only response was to laugh along. "Graduate student" was not a status all elders readily recognized or had a name for anyway, except *nyathi e skul — skul maduong'* (a schoolchild — in a big school). A schoolboy arriving with no shorts but long trousers, and at times a motor vehicle, was an oddity enough to begin with. (Revisiting years later, once teaching in another university, would be easier to explain.) This odd pupil seemed to ask at times about things too obvious or far-fetched even to mention. Sometimes, by laughing at first, my interlocutors let me know there was something deep and unspoken to ask about, to listen to more carefully, maybe to think about some radically different way.

Among those topics were the debts and obligations involved in mating and marriage, and in life and in death. On first arrival I already knew that credit, debt, and claims over land and graves were things I wished to look into. These had to connect to each other and to much else, my intuition told me, but I had hardly the dimmest idea how. It did not become clear to me until years later that much of what I had been studying, and collecting words in several languages to describe, would somehow fit under the broader heading of entrustment—one which, along with its reciprocal obligation, belonged to no one discipline.

The more I learned, of course, the more beholden to the people of the lake basin, and others farther away, I became in my own way. They were entrusting me, and those working with me, with a way of life and with their current conditions. They were passing me a ball, to pass along abroad (as some explicitly asked), to speak and write about. They were passing it too for their grandchildren. But it would be wrong to have illusions about this either. They had their own oral traditions no less important, and their own novelists and scholars of accomplishment one could only admire. The ball would need to be passed back and forth. And anyway I was not the first person from outside to try to describe the place or what went on there.

A Cryptic Remark

A British district commissioner wrote in 1909 from the hills above the lake, that sea-like source of the Nile he knew as Victoria. G. A. S. Northcote was one of the first to set up administrative stations in the region, and he had now been there several years among the Luo people (or to him, "Nilotic Kavirondo") and others including Gusii, Suba, and Kuria (to him, still grouped with others as "Bantu Kavirondo"). In his annual report, in the middle of the chapter on "Native Organization" he noted, under the combined headings assigned by protocol as "Debt," "Beggars," and "Slander and Abuse," just a single conflated sentence: "No particular regulations govern these matters: beggars do not exist: owing to the multitude of unsettled claims the individuals of all these tribes may be said to have been in a state of chronic indebtedness."[3]

Chronic indebtedness. Certainly others, from inside and outside the area, have used words like these since. But Northcote's cryptic sentence begs some questions. If no beggars—yet—then indebtedness to whom? For what, and why? There was as yet little public contact with banks, landlords, or a company store. Land rental was rare and culturally disapproved. Few "natives" yet worked for foreign firms in the district, and mining firms were just staking their first claims for digging. Marketplaces were still few and far between. Minted money had been in general circulation for only a few years, and most of the economy took other forms. Taxation had begun but was not yet the chronic headache it would become. Did "chronic indebtedness" necessarily mean destitution? Not necessarily, since one can live rich in the red and just pass all debts to heirs. Need it mean unhappiness? No again, since a person in debt might derive much satisfaction from story, song, or dance. Must it mean incapacity? No once more, since prophets and politicians with bills to pay have shaken the world. Why was this section of the report so short, when other sections, like the ones on inheritance or succession of "chiefs," went on for pages? Was it perhaps that prescribed headings like "debt" or "beggars" made little sense in local context? Or was it instead that there would be too much to say?

The pages that follow are about ways of understanding credit and debt, and about people in an equatorial setting who seem to live in chronic indebtedness—yet not necessarily destitute, unhappy, disabled, or without credits to their name. To be sure, some among the Luo—the largest of the "tribes" in Northcote's report, and a group since multiplied severalfold to some three million souls—can indeed be described as poor, miserable, or incapacitated, for reasons that clearly include indebtedness. But the Luo country and western Kenya are not a land without joy, and it would be wrong to infer that life would turn rosy if all debts and obligations were cleared. For some of those debts that never go away are the very fibers that hold society together and give it form and flavor to those who live in it.

Let us look, then, at what "indebtedness" is made of, and how it ties together the people it does. To do this we must be willing to cross the line freely between what looks secular and what looks sacred, exploring not just bank loans and corn seed programs but also marriage payments and sacrificial prayer.

But what, the reader might well ask, do all these things have to do with each other? This will take some explaining. It will help to resurrect an obscure old term like "entrustment," one broader and deeper than "credit," but woollier too. Some of the explaining will need to be done in terms with inverted meanings (did you ever hear "creditworthy" refer to a lender?). Some will require separating our conventional terms as if by centrifuge: perceiving, for instance, that two kinds of "obligation"—to oblige and to obligate—while being sister terms from the same Latin root can yet refer to quite different, even opposite things. Entrustment and obligation, framing "indebtedness" in a less cold and negative cast, are the two central English terms in this volume.

Not least important, the explanation requires making an attempt to see the world the way Luo do. English, and even Latin, take us only so far, and then we must look to indigenous African terms and concepts. One Luo word, *holo*, for instance, means to borrow *or* to lend. Another, *agok*, means both "shoulder" and "diverted repayment." To appreciate how Luo people communicate, we must also note how they "switch codes" between their own Nilotic tongue Luo, the trade language Swahili, the administrative language English . . . or language left unspoken.

For the moment, suffice it to note just a few key points. First, to speak loosely, not all kinds of "accounts" in Luoland—or anywhere like it—flow freely into each other. A farmer can be poor in the corn seed account but rich in the land or the marriage cattle account, or poor in all these but rich in the sacrificial account (if one may speak of such a thing at all). Second, not all the flows between the different sorts of "accounts" are two-way. The corn seed account cannot be readily replenished by tapping into the bridewealth account, for instance, but the bridewealth account can (and often does) get siphoned out of the corn seed account. Third, never have all debts been settled, nor will they ever be. Personal, family, or community debts, like national debts, seldom really get squared up in their totality. Fourth, this may not be such a bad thing—for Luo, or for their creditors near or far. There are some debts Luo people want desperately to get rid of, but others they desperately need. And there are creditors (in the World Bank, for instance) who become quite worried (in this case, for their own well-being) when people in places like equatorial Africa do not borrow.

Finally, some kinds of "accounts" serve as models for other kinds. The accounts in the here and now, the corn seed or money accounts, for instance, serve as models for the accounts in some sort of "hereafter" (as in one variant of the Lord's Prayer—known to many Luo and surely to Northcote— "Forgive us our debts as we forgive those who debt against us"). Much of economic life is symbolic, much of practical life metaphoric. Beyond a certain point, though, it may no longer make sense to think in terms of "accounts" at all, because accounts demand to be settled, or equalized, in ways that not all the rest of life—or life with death as an accepted part of it—does. Sometimes we can discharge a debt only by passing along something to someone else who was not part of it in the beginning.

What is true for the Luo may be true for all of us. We live in perpetual indebtedness, but also perpetual "increditedness." Seldom do the two balance out to zero. Many of our debts make us unhappy, but not all of them. Not all our credits can be used to pay off our other debts, because our minds partition the world into compartments, some connected only by one-way siphons or none at all. Nor will the water ever come to equilibrium—unless it be at life's end or somewhere beyond our ken.

A Way of Starting Over

It is no secret that the history of international interventions in the agrarian African tropics has not been a happy one, and that one of its main legacies has been debts beyond hope of repayment. The past century's rocky relations between Luo and their national and international financiers have made western Kenya a despair of developers, and these have usually ascribed the problems to "constraints" or "impediments" inherent in Luo culture. But the Luo are not an isolated people; few of the "development" problems encountered among them are unique to them or to western Kenya. Looking more sympathetically into what international financiers consider one of their "problem populations" (a sentiment not unrequited) illuminates not only a complex and fascinating culture misunderstood by donors and lenders. It reveals as many cultural constraints and impediments in financial centers as in the hinterland, implying a need to rethink rural development and its finance from the start.

Rather than dwelling only on transplanted institutions and their shortcomings, and smugly concluding that rural Africa is a hopeless mess, I prefer to take an opposite, more inductive approach. The first question to ask is what ideas and practices *do* seem to work in a given setting, or at least to persist there, and why. This approach means not confining our understanding of credit and debt to banks, cooperatives, statutory boards, and so on. I suggest extending the view to include various kinds of indigenous and imported practices I call entrustment—an idea one step more abstract, and a shade more inclusive, than credit.[4] Entrustment implies an obligation. This pair of ideas can cover borrowing and lending of non-economic as well as economic kinds. They imply the trust or confidence (in an individual or set of persons, or in a process or system) that underlies a transaction with a lapse of time.[5]

Definitions structure perceptions. Financiers who define credit as purely a financial or official concern have tended to conclude that rural East Africans, and Luo in particular, enjoy very little of it. Widen the definition just a bit, to mean exchanges expected to be reciprocated over time, and many kinds of transactions appear, in the East African hinterland as elsewhere.

Land, money, animals, labor, and even humans themselves are all transferred and returned later. But concepts like credit and debt themselves have their limits. In rural East Africa transfers and counter-transfers made over time are often not only hard to quantify and evaluate, but moreover hard to classify as economic or non-. Women exchanged for cattle in gradual marriage payments, land borrowed and lent with ancestral graves on it, or labor contributed to funerals are economic, but never *just* economic transfers. They are credit and debt in some aspects, but not in others. Another terminology is needed.

By taking into consideration not just credit and debt but entrustment and obligation more broadly, we obviate the burden of trying artificially to sort gifts from commodities, or the economic from the non-. Entrustment implies an obligation, but not necessarily an obligation to repay like with like, as a loan might imply. Whether an entrustment or transfer is returnable in kind or in radically different form—be it economic, political, symbolic, or some mixture of these—is a matter of cultural context and strategy.

Nor need the recipient necessarily repay the giver in all cases. Some obligations may be discharged by giving or passing on a benefit to someone else, in series. So it is with inheritance, in which, often, by receiving something from a member of a senior generation—a plot of land, piece of crockery, or a ring—one accepts a shared understanding that it is an heirloom to be passed "down" through a longer line of kin or somehow shared with others in a group. Asymmetrical or serial transfers and exchanges like these, involving multiple generations, have received far less attention by anthropologists and others in African and tropical settings around the world than have, say, gifts; but they can number among life's most meaningful, even sacred, acts and responsibilities.

With this entrustment approach, a great variety of activities spring into view as the cultural-economic backdrop for borrowing and lending, including "formal" credit and debt. We become better able to understand how commitments that are not symbolically comparable may nonetheless conflict in an economic way: how, for instance, an obligation to a rotating savings club can be used as a defense against obligations to an acquisitive spouse, or how an obligation to an ancestor can clash with obligations to a state credit bank. We come to see how the kinds of thought and action that economists class as "moral hazard"—that is, defaulting despite ability to repay—may have their own kinds of morality at their base. If economic capital can be transformed into "symbolic capital" and vice versa, then entrustment and obligation are likely to prove a more accommodating pair of concepts than credit and debt.[6] They befit a context with few clear lines between economic and non-economic, between material and nonmaterial, between exchangeable and unexchangeable.

Entrustment and obligation are not just practical matters but moral and sometimes aesthetic ones as well. Strategies involved in these dealings are based not just on economic self-interest as much as on the satisfaction of different social or cultural norms—for example, in projecting an image of oneself or creating a proper (not just profitable) marriage, funeral, or sacrifice. Reason and rationality pertain not just to individuals but also to groups, networks, and categories; and they do not always involve measurable gains. Nor are reason and rationality the only human aspirations. Too often they leave out the intuition, the experience, the feel.

Chapter Outline

This much should give a taste of what I mean by entrustment, obligation, and belonging, and some bearings for the part of the world to be considered. Entrustment can mean more things than I could possibly attempt to cover (particularly if including, say, entrusting a secret to a friend, or justice to a court). So might obligation. Here we shall be concerned with the kind that has a material dimension: where it is actual persons or things— animal, vegetable, or mineral, alive or dead—that are entrusted or need to be returned or passed along. And while I try not to neglect movement between town and country, the focus will be on the rural places, where contact with the tangibles can be so direct, and on the closer ties there.

Chapter 2 presents some anthropological and other approaches to credit and debt, and more broadly, to what I call entrustment and obligation, in roughly chronological order of appearance in the literature. Social and cultural anthropologists, with their often relativistic and holistic views, have tried hard to find reason and meaning in local practices of exchange, and to describe the groups, networks, and categories of persons involved according to local understandings. But far less attention has been paid to the loan than to the gift or the sale; credit and debt have usually appeared as incidental to other topics. These concepts translate many ways between tongues, and intimacy and familiarity are critical concerns. An approach to economic culture that combines the advantages of absolutist and relativist perspectives is likely to work best. Ethnographic examples in this chapter, drawn here from around the world, help set the Kenyan Luo case into context.

The study turns next to that case. A brief introduction to the Luo country, in Chapter 3, describes the landscape, the varied ways its people make a living, and ways they get along with others. The activities most important to Luo economy, like grain farming, are not always the most highly valued culturally or necessarily the ones Luo people deem the best uses for "agricultural" finance or reinvested farming profits. Livestock, education, and opportunities to migrate and work in cities or abroad are much favored, as are funerary and other ceremonial expenditures that have hitherto meant less to official economic planners. Crops, livestock, and other ways of making a living are strongly tied to particular genders or ages. Often, then, decisions about allocating loans or finding repayments mean choosing between conflicting interests of different members of a domestic group. Luo people have their own subtle, diffused, and in some ways sophisticated system of political organization, not always compatible with state authority. They appear profoundly ambivalent about a universal currency under government control. Not everything is freely exchangeable for everything else, and not everything for money. Commodities, like people, have histories, and time is a central part of decisions about who can exchange what with whom, and why. The chapter closes with a brief discussion of methods used in research, and of terms in the Luo language that offer insight into ways people think about fiduciary dealings.

The remaining chapters of this volume survey, one after another, some things that rural Luo and neighboring people borrow and lend, or entrust, locally among themselves. Not just inanimate "goods and services," that is, but also living and breathing beings are also "lent" around. We begin in Chapter 4 with what a Luo experiences in early childhood. Small children themselves are objects of entrustment, cared for by other, older children from inside or outside their homes. Children and their caregivers move between country homes, and sometimes between town and country. Cattle and other livestock, similarly, are entrusted or "driven" from one home's herd to another's, for safekeeping that may continue for decades, as a way of evening out imbalances in homestead resource endowments, concealing wealth, and marking special friendship. At times, it can happen that cattle and other animals residing in a homestead on loan are tended by children and youth who are there only on a kind of loan too.

In Chapter 5 we look into entrustment of labor, tools, food, and shelter. In one way or another, Luo people borrow and lend practically everything, and what one borrows or lends helps define who one is. A people often described as egalitarian turn out in some ways to be notably hierarchical, and entrustment and obligations play a big role in social status. Short-term rotating labor associations and seasonal share contracting show the innovativeness and flexibility of local arrangements for spreading out resources unevenly divided between families, but arrangements that appear equal on the surface do not always turn out so. Equivalencies and fairness appear to be calculated not just on the basis of economic or (more narrowly still) monetary values of goods and services lent, but also, sometimes, on the basic of symbolic correspondences and parities decidedly at variance with those calculations of worth. Loans fair in one way are often unfair in another.

Luo customs of entrustment and obligation are not atavistic residues of some pristine African past. They are norms and habits continually adjusted. Various sorts of contribution clubs arise locally, with or without any help from outside, and they can include dozens of members. They contain valuable lessons about credit, saving, and sociability, showing that it is not just how one saves or borrows that matters but also how and when one appears to others to do so.

Not all important exchanges occur between peers or contemporaries. Entrustments and obligations involved in labor migration and school fee payment in the past century carry expectations that when the beneficiaries eventually make good, they will provide one or more of their junior kin in turn with temporary food and shelter in town, or supply school fees and expenses. This last sort of arrangement links not just one or two kin but an individual and a kin group as a whole, and the obligations involved can last lifelong. People living in the shadow of debts like these cannot be expected to consider impersonal debts to state cooperatives or banks their highest personal priorities.

The role of exchange and entrustment in marriage, something not always easy for foreigners to understand, is given some explaining in Chapter 6. What is basically a tradition of transferring wealth—in cattle, and more—from groom to bride's father, ostensibly in exchange for her, turns out to have a more complex back and forth, involving many other people on both sides. And the transfers or "payments" and visitations between groups are not made all at once but can be protracted over years, decades, or even generations. On separation or divorce, not only does the return of cattle become an issue but also the compensation for hospitality already consumed. What makes these ways hard for many to grasp is that not all "customary traditions" are practiced all the time. They come and go, and are continually debated; and eloping or simply forgoing formalities is an option seen more often at some times than others. In all these transfers, the calculation of worth involving both humans and animals can be hard, especially when the passage of time changes these economic and cultural values and makes it necessary to renegotiate old agreements. Between married and unmarried also lie shades of gray — child bride pledging, pseudo-marriage, sex trading, and so on — that thicken in lean times, raising questions of obligation in pregnancy and childbirth.

Chapters 7 to 9 complete the basic survey of Luo local entrustment by examining dealings between the living and the non-living, including spiritual beings and forces. Mortuary celebrations, discussed in Chapter 7, are usually done in two phases and are among the biggest and most elaborate events in Luo life. Not only do they involve contributions from kin and others, to whose funerals in turn the family of the dead are expected to contribute, but they also involve a reckoning of the debts of the dead. Redeeming lost lives in the here and now is an old Luo and central and eastern African tradition, but one in abeyance in recent history when state authority has taken over some of its function of social control.

Inheritance, the topic of Chapter 8, offers another window into some of the deepest Luo and East African values. The term includes the passing of possessions between the aging while still alive and their juniors (or devolution inter vivos). It is part of serial entrustment, providing material substance to the lineal continuities that are precious to Luo and others living around them. It has much protocol to follow, but circumstances often make that protocol hard to follow in a way that keeps everyone happy — for instance, when junior heirs have more children of their own to feed than do the senior ones who would ordinarily be given larger shares. How people handle hard cases like this tells something about their cultural programming.

Chapter 9 treats ways Luo people, like other Nilotic speakers, conceive of relations between themselves and their spirits and divinity as being, among other things, ties of entrustment and perhaps obligation, if not actually of credit and debt. Humans apply metaphors from their own worldly dealings to their dealings with these beings. Blood sacrifice, a traditional way of communicating with spirits, has been under siege by church missionaries who prefer their own sacrificial idiom and commemoration. But Luo and neighboring people in East Africa have blended older indigenous with newer Christian traditions of sacrifice in creative ways, both in short-lived prophetic movements of radical protest and purification and in longer-lived ones that have become established churches—both with apocalyptic visions.

The concluding chapter discusses indigenous forms of entrustment in the round. I suggest, among other things, that credit and debt, and entrustment and obligation, are not just economic but also cultural and symbolic, and sometimes political too; and that neither absolutist nor relativist perspectives are adequate on their own to understand their nature. Spheres of activity that seem unconnected on the surface turn out to interthread underneath, but not so tightly as to make causes and effects, reasons and responses, wholly orderly or predictable. Familiarity and social distance are critical, and there is no single way to measure these—particularly when the participants include those no longer living. To see what meaning has to do with wealth and power in exchanges conducted over time, we need to think in several ways at once.

A proper perspective on entrustment means moving anthropology beyond its obsession with "the gift" and reciprocity, and perhaps political economy beyond "the contract," the rational or self-interested actor, and the market and state. It involves broadening our language beyond our mother tongues to look for transactions, agreements, and metaphors that have no simple translations and may indeed be unnamed in any language. Understanding entrustment means understanding the reciprocal *and* the serial, the contemporaneous *and* the intergenerational, the secular *and* the sacred, in ways not yet roundly explored.

CHAPTER 2

Fiduciary Culture

A Thread in Anthropological Theory

What force is there in the thing given which compels the recipient to make a return?
—MARCEL MAUSS

Fiduciary culture—the culture of trust and entrustment—is only beginning to come into its own as a topic of study, but it has deep roots and noteworthy antecedents. Since well before Herbert Spencer's and Charles Darwin's time, and before anthropology came into being as a named discipline, the topic has played a part in debates about the evolution of civilization and about morality in economic life. Three questions have recurred in the debates, springing up in different guises and in various disciplines. One is whether barter, cash sale, and credit constitute any kind of evolutionary sequence, one built on another. A second is whether economic institutions (in the sense of patterned ways of thinking and acting) naturally grow distinct from non-economic ones over time, and become ever more specialized, as part of a process of increasing complexity of society and a tightening interdependence of its parts. A third, normative question, linked to the first two, is whether everything ought to be made freely exchangeable for everything else: whether, for instance, land, labor, or livestock ought to be freely convertible, one to another, in markets—and if so, who ought to control the terms of exchange and the terms of discourse. How one approaches these three questions has profound implications for our understanding of human nature, human condition, and ways these vary.

This chapter outlines a few observations that anthropologists and others have offered on what I am calling fiduciary culture. Anthropologists have always had insights into credit and debt, but they have seldom placed these at the center of their studies the way economists have. Mostly they have written about borrowing and lending, and about other forms of entrustment like inheritance, only in passing, in works on other topics. The political-economic threads have crossed with the symbolic ones, but these have seldom been tied. The study of fiduciary culture has been a story of explorations and discoveries never quite consummated, of an intellectual itch never really scratched.

The Gift, the Loan, and the Sense of Obligation

The turn of the twentieth century already saw strong anthropological challenges to evolutionary theories that held that societies and economies progress from barter to cash to credit. Franz Boas, much helped by his local collaborator George Hunt, perceived credit as being at the heart of the ostentatious gifting in the potlatch of the Kwakiutl (or Kwakwaka'wakw), Haida, Tlingit, and neighboring coastal peoples of the Canadian and Alaskan Pacific.[1] Just after the first "world" war, Bronislaw Malinowski (1922) described, in the Trobriand and broader Melanesian archipelago *kula*, an elaborate circuit of delayed reciprocities, in enduring partnerships, that occurred in parallel with immediate market exchanges, and that tied tightly into kinship, leadership, myth, and magic, among other things.

Marcel Mauss, a nephew and protégé of Emile Durkheim who also followed Boas and Malinowski in the early 1920s, noted that all societies have "customary" forms of credit, whether monetized or not. Mauss eschewed the "unconscious sociology" of the commonly imagined barter-to-cash-to-credit sequence, but he was enough of an evolutionist still to want to speculate about origins and developmental process. Mauss wrote that the origin of credit is to be found in the gift, especially in "its ancient form of total prestation." Out of gifts given and later returned arose barter: the time lag simply shrank to disappearing. Also out of such gifts and counter-gifts arose purchase and sale, and the loan. Indeed, the gift "necessarily implies the notion of credit . . . there is nothing to suggest that any economic system which has passed through the phase we are describing was ignorant of the idea of credit, of which all archaic societies around us are aware" (Mauss 1967, 34–35). The origins of these things are unknowable, in my opinion; but that means there is nothing to disprove Mauss on the prehistoric sequence; and I know of no society, in his time or our own, where gifts and loans cannot both be observed.

While acknowledging that gifts and return gifts are basically a form of credit, Mauss concentrated more on gifts than on loans per se. He argued that gifting is not voluntary and disinterested, as so often assumed, but obligatory and self-interested. On what made it compulsory, Mauss followed Maori and others to an area beyond the figurative, waxing even into the mystical: the gift contained something of the giver, and it could potentially damage or "poison" the receiver who did not respond properly by reciprocating or passing on the favor after an interval. "Charity wounds him who receives," he also noted (p. 63): words from which patronizing aid agents could well learn today. More broadly, what we all must learn from "archaic" or primitive people, argued Mauss, is that every human has three obligations: to give, to receive, and to reciprocate. What matters is not to gain things but to forge and maintain social relationships—to Mauss the most important thing. With or without the mysticism of person transubstantiating into gift, or of giftly poisoning, Mauss's point about the social importance of giving things away and keeping them in circulation, not just accumulating them, was a compelling challenge to orthodox economic thought. It established "the gift" as a central topic of economic anthropology.

Further ethnographic support for the point would not be long in coming. Cora Du Bois (1969, 48) noted on Alor Island (in the Netherlands East Indies, now part of Indonesia), in the 1930s, that people sought to get rid of, or conceal, wealth received (here, gong currency) just as soon as possible after receiving it—if not by lending it out casually or putting it into feasts of competitive giving, then at least by burying it to conceal possession. Pigs received were immediately farmed out to stock-care partners in other villages. "People thought my kinsman Lanmani was poor because he did not have a lot of outstanding debts," said Fantan of Alor to Du Bois (Du Bois 1960, 150).[2] Here, as in East Africa, it is meaningless to try to survey or calculate wealth by totting up possessions without also finding out what people are owed. East African animal keepers, like Alorese, lend or entrust their large animals widely to special friends, for reasons including, among others, concealing ownership and cementing social bonds. In the Pacific or in Africa, this way of turning tangible wealth into intangible obligations has seemed to many analysts precisely the opposite of a commoditized, market economy. So it may be, in a sense, but it is important not to forget that the *same people* can—and probably everywhere do—participate in both.

Like his mentor Durkheim, Mauss looked upon economic life among "primitive" or "archaic" peoples—somewhat dubiously, he likened one to the other—as being driven mainly by social and moral concerns. Credit, to Mauss, is only an elaboration of the gift, and the gift is not just economic. In "early" societies, Mauss wrote, exchange was never merely an economic institution but simultaneously social, religious, moral, legal, and aesthetic too: something more total (Mauss 1967, 1). Markets were "familiar to every known society" (p. 2), even antedating money and merchants. Extending Durkheim's understanding of social facts and the importance of social solidarity, Mauss claimed that "Exchange, whether by gift, barter, or credit, had to do mainly with groups, not individuals: For it is groups, and not individuals, which carry on exchange, make contracts, and are bound by obligations; the persons represented in the contracts are moral persons — clans, tribes, and families" and they form part of a "system of total prestations" (Mauss 1967, 3). They exchange not just material wealth, but also intangibles: "courtesies, entertainments, ritual, military assistance, women, children, dances, and feasts; and the fairs in which the market is but one element and the circulation of wealth but one part of a wide and enduring contract" (Mauss 1967, 3). Mauss, like Durkheim, sometimes exoticized the "primitive." He also stereotyped (or "occidentalized") the modern "West" as being colder, more individualistic, than it necessarily is. (For a counterexample, one might point to a symbolically salient and formative time like Easter, when Euro-American parents always somehow see to it that no child in the egg hunt scramble is left without eggs in the end.) Mauss considered social history as a matter of individuation.[3] If it went too far, he concluded, and civilization forgot the importance of gifts, obligations, and ties between groups, the world would face more experiences like the Great War that Mauss had just lived through.

While many in the humanities and social sciences have taken Mauss on the gift as the first and last word, others equally inspired by him, even within the French anthropological circle, have subjected it to criticism and reanalysis.[4] Alain Testart, for instance, has noted that gifts and loans vary greatly from one context to another in just *how* obligatory they are in their return, and in whether failure to do so implicates the receiver in any loss of status or legal trouble. He traces several cases on a scale between ones like charitable donations, where (in his view) no obligation to reciprocate might be felt, and at the other extreme ones where, as "in numerous precolonial African societies, the creditor can seize the person of the insolvent debtor and make that person a slave."[5] He finds, though, not an unbroken continuum but two groups: those in which a return cannot be required and demanded by the donor or any agent (gifts, strictly speaking) and those in which it can (credit). The northeast Pacific coast-and-island potlatch (where there is a social but no legal sanction for giving back) and the southwest Pacific inter-island kula (where a donor or "creditor" may seize a kula object to get even, but not harm the recipient's or debtor's body) fall close to the middle. Between the two broad types of exchange Testart perceives, there are surely many settings and situations where a donor might not demand that a repayment be made but kith or kin of the donor or recipient around do, if only for their name or reputation.

The Luo case calls for a less doctrinaire and perhaps more complex understanding than Mauss's, and a less rigid separation of "archaic" and "modern" societies. It has always been individuals *and* groups that engage in credit and other exchange. Often when individuals borrow and lend, there are kin groups or other groups behind them with strong interests in their transaction. Perhaps more importantly, individual interests have always hidden behind group interests. Society appears never to have known a time of overall contractual solidarity such as Mauss imagines, nor does it ever really individuate into what he stereotypes as an occidental condition of atomized individuals.

Less contestable is Mauss's point that wealth can take the form of people, not just things: a concept that remained influential in African economic anthropology to the end of the twentieth century (Guyer 1995) and into the next. While it is important not to consider people or animals as *only* wealth, Mauss's insight points the way to questions only now being asked: what the continuing, adapting African institutions of bridewealth (that is, marriage dues) and human pledging have to do with fiduciary finance from afar. The answer, of course, will be more than a little.

For all the importance of his comparative work, Mauss was missing something. The gifting he treated took place between living contemporaries. He had little to say about bestowal and obligation between members of different generations. Had he studied Africa, he might have directed more attention to this gifting turned vertical, and to the interesting questions it raises about obligations between the living, the dead, and the unborn—and of how these might relate to obligations between the living themselves. These are important African concerns, and as I attempt to show in chapters to follow, they are ones on which Luo and other people in equatorial Africa have wisdom, opinions, and experience to offer.

Fifteen years after Mauss published that crucial little book, Ruth Benedict, an American student of Boas, picked up on the theme of reciprocity and obligation, and she plugged it more directly into the affairs of war and peace, such as Mauss had speculated upon in his conclusion. But by then there was a new intercontinental war to talk about.

Studying Japan from across the Pacific, on behalf of the U.S. government, Ruth Benedict found obligation to be a key to Japanese culture, and a way to crack the paradoxes and mysteries that had so perplexed her countrymen. She sought to explain, for instance, the seemingly infinite loyalty and bravery of kamikaze suicide pilots as well as the docility of Japanese prisoners of war who seemed to deem themselves disgraced and dead as Japanese, and reborn as something else (Benedict 1989). A sense of indebtedness and obligation (*on*), learned early in life through maternal and other attention (in child bathing, schooling, hazing, and so on), elided into obligation toward teachers, parents, ancestors, emperor, and empire. Repayments toward never-ending debt (*gimu*) were conceptually distinct, though, from repayments toward time-limited, more easily dischargeable debts (*giri*) to one's creditors, the world, or one's blemished name.

Like much of Benedict's writing, her totalizing portrait of Japanese obligation-obsession as a "configuration" or "pattern of culture" ("For in Japan the constant goal is honor" [p. 171]) could be and has often been criticized as oversimplification or cultural caricature. And yet few serious Japanologists, inside or outside that country, would say it is not trenchant. One lesson to draw from it is that we humans can feel as much obligation to the dead (and unborn) as we do to the living: that big topic Mauss had tended to undertreat. Another is the difficulty of translating one culture's terms for debt or obligation (and maybe therefore for credit or favor) into another's.

Kenyan and other East African people could be said to share with Japanese a weighty sense of intergenerational indebtedness. In both countries it can give rise, in youth, to extremely deferential behavior and explosive rebelliousness. Yet the differences are great too. Japanese nationalist and imperialist sentiments find scant parallel in equatorial Africa, where race, language, and nation are usually deemed to coincide less neatly, and where the carvers of nations, as now delineated, were anyway Europeans, not Africans.[6] Nor does one often find in Kenya the kind of lifelong loyalty between company and worker for which Japan became so famous in the postwar decades—a difference that must have affected economic outcomes somehow.[7] That the Kenyan and Japanese cases compare closely in some cultural respects but contrast sharply in others has implications of its own. It casts serious doubt on whether any society conforms, or can for long, to a single organizing pattern, a master theme or mold. Indebtedness and obligation can be found everywhere. But every duty conflicts with other duties (as Durkheim noted), and duties change.

Exchange Spheres and Time

By the early to mid 1940s, as Benedict was exploring the cultural psychology of Japanese people and their senses of indebtedness both specific and general, the economic historian Karl Polanyi was busy criticizing Euro-American understandings of markets and exchange as they had changed over the centuries. In his view (doubtless partly inspired by Marx's on the fetishism of commodities, in *Capital*), they had become more unrealistic. Since the industrial revolution, land, labor, and money had been spoken and written of as commodities in markets. But they were *not* just commodities. They were vital aspects tied in with the rest of human life, and to abstract them and deem them subject to market forces alone, free of any secular or sacred social regulation, could only do injustice to the real people concerned, denying both nature and human substance (1957, 68). Hence he and his followers created the culturally relativistic school of thought known as "substantivism." Substantivists opposed the absolutism and "formalism" of neoclassical economics with all the latter's assumptions about self-interested ("rational") individuals, supply and demand, and free choice.[8] As Polanyi saw it, unregulated European bankers had overextended financial credit overseas after the First World War. They thus helped bring about the Great Depression of the 1930s, and to drag into it the people wherever Europeans had extended loans. And that included Africa.

The anthropologists who followed Polanyi's critical, relativistic way of thinking after the Second World War, and applied and adapted it in Africa, were more concerned with local than international economic life. General ethnographers deeply immersed in "field" study using local languages were fascinated by particular indigenous commodities, noncommodities, and quasi-commodities. Those working in the Pacific and in Africa influenced each other back and forth in this period, still in the heyday of functionalist anthropology. Paul Bohannan, an American doctoral student of Edward Evans-Pritchard's at Oxford, was inspired by Karl Polanyi, on exchange, markets, and market fictions, and also by Malinowski, Raymond Firth, and others who had studied exchange circuits in Pacific islands. Malinowski and Firth had both found items exchanged for only specific other items. In the Trobriand kula case in Melanesia, Malinowski had found white shell armbands exchanged only for red shell necklaces, neither having any utilitarian value other than exchange or display. Firth, on tiny Tikopia island in Polynesia, had similarly found separated spheres of exchangeability for subsistence and luxury goods.

Now Bohannan (1955, 1959), studying Tiv people in northeastern Nigeria, came up with an African variant. Tiv kept separate, ranked exchange spheres for what he called "subsistence" (food, utensils), "prestige" (slaves, brass rods, and special cloth), and "marriage" wealth (non-slave women). They sought to convert wealth upward toward marriage and reproduction but not downward toward subsistence, since that would show weakness and desperation. Not everything, then, was freely exchangeable for everything else, once culture was properly taken into account; but cash and markets threatened to make it so, scrambling distinct "spheres." Fredrik Barth (1967) built in turn on Bohannan's observations in his own research among Fur people in Sudan, showing individual entrepreneurs becoming wealthy by manipulating or contravening the local rules and conventions about "spheres of exchange" and moral barriers between.

In the 1980s and '90s, other anthropologists working among Nilotic people (Shipton 1989 on Luo, Hutchinson 1996 on Nuer) found that money itself—far from being a universal exchangeable—was itself locally divided into separate spheres of exchange with different moral valences. In the Luo case, for instance, cash deriving from socially approved activities and exchanges could be used for special purposes like bridewealth or funerals, but cash from unapproved activities (like selling patrimonial land, or gold or cannabis) could not be used for these without spiritual dangers to self, family, or lineage, or without expensive and risky purification ceremonies. Anthropologists elsewhere in Africa (Ferguson 1992, Piot 1991) and the world (Parry and Bloch 1989, Zelizer 1994) were similarly finding that wealth was divided into different categories for short- and long-term uses, and that money was "earmarked" for varied purposes with differing moral casts, not always connoting something immoral, antisocial, or individualistic. In the Luo country in Kenya, daughters are not ordinarily pledged for grain; but in famine times, fathers in families desperate for food have pledged or married them off for food without evidently incurring moral opprobrium from their neighbors. Jane Guyer (2004), restudying Tiv and some of their Nigerian and Cameroonian neighbors, found strong indications that "spheres of exchange" (or seen the other way, rules or expectations about nonconvertibility) mutate in systematic, maybe predictable ways over time and space. More is said about exchange spheres in chapters to follow as we look further into plow team sharing, bridewealth, funerals, inheritance, and sacrifice. The cognitive partitioning of economies into discrete imagined spheres or "sectors" is no less a feature of thought in international aid agencies and their offices in capital cities than in the kinds of societies that Firth and his contemporaries called "primitive."

Substantivists and formalists both had their points, and their debate has lapsed into stalemate, though it has sometimes been rekindled in different terms. Substantivists tended to devote more attention to gifts and ceremonies, formalists to sales, markets, and prices; and often they talked past each other, as anthropologists and economists have done more generally. In truth, it is hard to find any real people who represent a "gift economy" or a "commodity economy" pure and simple (see Gregory 1982). The no-man's-land in between, which neither camp ever really covered as a cultural and economic practice, is the loan.

Enough had been written, by the 1960s at least, to know that even small-scale societies had lending and borrowing, and that a loan or credit need not mean just money or market principles. The one important attempt to sum up the picture across cultures and economies, a collection published by anthropologist Raymond Firth and economist Brian Yamey in 1964, presented it like this: "Even in the most primitive non-monetary economic system the concept of credit exists — the lending of goods or services without immediate return against the promise of future repayment. It involves an obligation by the borrower to make a return and confidence by the lender in the borrower's good faith and ability to repay. It may be equivalent in value to the loan or augmented in value above the loan (i.e. with interest). The augmentation may be voluntary or prescribed, and it may be proportionate or not to the amount of time for which the object lent has been held. The repayment may be contractual and enforceable at law, or it may have no legal backing but be socially binding" (p. 29). The volume contained cases from all over the world, except Africa.

Most of their summary would hold true enough there too (that is, if we can forgive or forget the authors' terms like *primitive*, not necessarily meant to condescend, and if we acknowledge that there are few places in Africa without money). If we add animals, and humans too, among things that can be lent or borrowed, and we include the dead and unborn along with the living, then we strain conventional economic terms like "goods and services," "credit," and "repay" but remain comfortably within the broader realm of entrustment and obligation. Some forms of transfer to consider, in Africa or anywhere else, are intimately familial in nature. And as Mauss surely appreciated, some are no more economic than symbolic, religious, or political.

As all this suggests, it can be hard to translate one people's terms for loans and exchange, or for agreements about them, into another people's mother tongue. Whether the English legal idiom can adequately capture the spirit of African debts and obligations was the subject of some academic pugilism in the 1950s and '60s. In one corner was Max Gluckman (1965) and in the other Paul Bohannan (1989). To both Gluckman, studying Barotse in Northern Rhodesia (now Zambia), and Bohannan studying Tiv in northeastern Nigeria, indebtedness appeared to be a central principle underlying social structure and cohesion. Gluckman, an anthropologist with some legal training too, found not just loan agreements but also damages and interpersonal harms understood this way, and he saw fit to assimilate these understandings to what he knew of English-style property, contracts, and torts, with some twists (for instance, *caveat vendor*, more important among Barotse than *caveat emptor*). But to Bohannan—here again the committed cultural relativist—a culture's concept of debt can be properly understood only in its own terms. To hammer these into any foreign legal categories—let alone Latinate legal ones, with all their archaic conceptual baggage—seemed to him an intellectual travesty. For one thing, doing so smuggled in misleading notions of private property. For another, it imported misfitting ideas about jurisdiction. Bohannan found among Tiv a lively tradition of moots, local informal hearings, whose aim was more to stitch together the torn social fabric by reconciling aggrieved parties than to punish them as infractors. Just as new money, to him, scrambled spheres of exchange, so new British-style courts only added confusion by layering over but not expunging basic Tiv understandings about how to right wrongs. Certainly, among Luo and their neighbors in Kenya, just as for Tiv, the colonial imposition of courts has added a new venue for addressing grievances, but the imported legal concepts may also have inflected, or hybridized with, what were once local understandings too.

If anthropologists have borrowed ideas about "spheres of exchange" from the Pacific in framing their questions about Africa, and vice versa, they have similarly transplanted other ideas about intimacy, social distance, and their bearing on exchange. A central work on this topic for anthropologists, attempting to systematize reciprocity, has been a volume by Marshall Sahlins, a specialist on Hawai'i and Polynesia, titled *Stone Age Economics* (1974), containing his 1968 essay "On the Sociology of Primitive Exchange." In this he draws a continuum between "generalized reciprocity" (or relatively free sharing or lending without strict accounting), through "balanced reciprocity" (tit-for-tat equivalency, strictly calculated and promptly reciprocated), to "negative reciprocity" (trying to get something for nothing out of someone, for instance by haggling, deceit, or extortion: in short, a rip-off). He overlays this upon a bull's-eye design representing circles of intimacy, from closest kin and neighbors to total strangers, after a fashion made famous by Edward Evans-Pritchard's 1940 ethnography on Nuer farmer-herder-fishers in Sudan: among other things a study of social and political relativity. Hypothesizing that intimacy comes with "generalized reciprocity," and strangerhood with "negative reciprocity," Sahlins reviews ethnographic evidence from dozens of societies in the Pacific and elsewhere, and finds that the theory holds up pretty well (pp. 193–96).

While Sahlins, like most of his anthropologist forebears, pays more attention to gifts than to loans, one can easily accommodate the latter to his schema. On the surface, one expects intimates to lend on softer terms than strangers: less likely to charge interest, for instance, less likely to insist on loan collateral as security, or more likely to forgive delays in repayment. In English, warmth and coolness, hardness and softness, are standard ways of describing the difference in feel between intimate and socially distant relationships. Evidence from rural Gambia, West Africa, tends to support the theory that easier lending, at least among Gambians themselves, tends to come with intimate social ties (Shipton 1990a, 1991, 1992a). And the same seems the case in western Kenya.

But matching up a linear continuum with a set of circular rings, and arraying exchange types and styles along it, risks losing the flavor and texture of the exchanges, the kinds one can not really measure. Even "warmth" has texture. Is it somber or playful warmth, concerted or haphazard? Is "coolness" formal, dutiful, boastful . . . or jealous, or spontaneous? In drying out exchanges for taxonomy, one risks overlooking *friendship:* the elusive, interpersonal chemistry of optative bonding that seldom seems to follow rules. Moreover, it is the *exceptions* to rules and expectations about reciprocity that are culturally most interesting. This is why the northern Pacific coast potlatch and the Melanesian kula are so interesting and famous: they involve long-distance exchanges (and in kula, enduring partnerships) between persons unrelated, but the essence of their exchange is gifts that demand no immediate, direct, or precise repayment, let alone theft or the like.[9]

Luo and their neighbors in East Africa, as we see in the pages that follow, are at times profit-seeking marketeers and at times reciprocators and redistributors. Their entrustments and obligations conform to no one such stereotyped logic. Nor is it easy to say that they are moving (let alone "progressing") from gifting to marketing, from generalized to negative reciprocity, or from socialism to individualism. For the same people who pay bridewealth, contribute to funerals, or share out prescribed portions of sacrificial meat also, in different times and places, squabble over bus fares, trade smuggled gold, or speculate profitably in urban land. Luo sometimes even engage in such very different forms of exchange with the same people.

Not just the people but also things can shift back and forth between "gift" and "commodity" exchange.[10] When, in the Luo country, Esther Ouma and her daughter distill *chang'aa* alcohol in their semisecret home distillery, they take grain from their family food store and convert it into a commodity for sale. And then, once Sylvester Nyatiti at the afternoon revel buys a glassful or a round of it to share with his neighbors and a stranger on hand, his handout, which may or may not be returned in kind, plays its small part in a network of reciprocities. If he later buys from Esther Ouma's father some cattle for bridewealth, with cash, he then exchanges these more as a gift that cements a relationship between groups than as the price of a bride. Once received they may well be sold again—very likely used in turn for the buyer's or buyer's son's own bridewealth—or entrusted to a distant friend for long-term pasturage and safekeeping. In the meantime they may well be stolen or swapped. Here there is no systemic "gift economy" or "commodity economy" but rather exchanges that weave in and out of these principles.

Entrustment and Classification

Human minds, wherever found, often seem to try to tidy up complex or messy pictures with simple categories. This process of classification, a main preoccupation of French anthropology for at least a century, applies no less to credit and debt than to anything else.

Dichotomization has been a main, career-long theme of the work of Claude Lévi-Strauss, an anthropologist strongly influenced by his French countrymen Durkheim and Mauss, among others, and best known for his comparative studies of mythic narrative. In his 1940s work on kinship and marriage (1969), Lévi-Strauss elaborated upon an older finding of James Frazer, Edward Westermarck, and others, that humans everywhere draw a fundamental distinction between "us" and "them" and impose incest avoidance rules on themselves that ensure they marry outside their kin groups. He started by observing that humans do not just exchange in dyads, group A with group B, period. Instead, it often happens that group A lets its women go to group B in expectation that it will receive women in turn from some other group C. Or the flow may change direction from one generation to the next. The whole system is based on trust. It must be, since women are more important than anything else a group might receive in return for them, the group may perish if it does not get enough women back, and groups yet do indeed let their women go in marriage.[11] Exchange of women (between groups of males, or of males *and* females) becomes the basis for political alliances between potential enemies, and indeed a basis of the commerce that supports civilization itself. And it all comes from the simplifying distinctions all humans make, like us/them. Whether women ought to be deemed exchangeable in the first place has of course been hotly contested since feminist anthropology bloomed in the 1970s. It may all be a figment of a few men's twisted scholarly imaginations. But it is likely that many people around the world do yet think of marriage as an exchange of humans, politically correct or not.

We humans not only dichotomize and categorize, Lévi-Strauss's huge opus suggests again and again, but we also align these cognitive distinctions with others. We take our distinctions from one domain of experience—you could say we "borrow" them—and superimpose them onto another to try to understand it. In this way (to take up from where Lévi-Strauss leaves off— for he gives little attention to economy or finance), we may talk of a loan, for instance, as warm or cold, soft or hard, as though these qualities came naturally with near or distant, respectively. We need not look far to find another example. When you've lent you are "up," "in the black," or "to the good"; when you've borrowed you are "down" ("in the hole") or "in the red," or maybe "to the bad." We use such language even though altitude, color, and goodness have no necessary correspondence with each other in nature. We break down our perceptions by simple distinctions, most often binary oppositions (often with liminal mediators of one sort or another), but then build them into complex structures by analogy and metaphor:

> credit / debt
> good / bad
> up / down
> red / black

Playful metaphorizing like this probably occurs in all languages. Often, though, the metaphoric structures are particular to contextual usage: "down" may not be associated with "red" outside of financial idiom, and in ordinary conversation, black can be associated with bad things (for example, black magic) as well as good. The semantic borrowings are seldom wholly systematic: the expression "in the hole" lacks an inverse like "on the hill." Nothing says human thought or speech need always be consistent.

Nor is there any limit to what can come into play. In Luo tradition land lenders, but not borrowers, may erect a symbolically potent, sharpened vertical spire stick (*osuri*) that points skyward atop a house roof, symbolizing maleness, seniority, and belonging all at once. That creditors or autochthons should be likened metaphorically to males, and debtors or allochthons therefore perhaps by inference to females, makes a statement about Luo cosmology and authority—and maybe about some hidden cognitive structuring propensity—that might be hard to discern in other ways. But findings like these also tempt us beyond our empirical data.

One of Lévi-Strauss's structuralist followers, the English anthropologist Edmund Leach, added a political dimension, and a sense of dynamism, scarcely found in Lévi-Strauss's own work. Leach observed in the 1950s in Pul Eliya, a farming community in Ceylon (later Sri Lanka), and later in highland Burma and in Britain, what so many others have seen before and since on the subcontinent, and what he might also have read in the biblical book of Proverbs (22:7): "The rich ruleth over the poor, and the borrower is servant to the lender." In one way or another, creditors always seemed to lord it over their debtors, and to get away with unequal exchanges. Social and political equality, and full or immediate economic reciprocity, were rare. So real social life was less about gifting than about a "network of indebtedness": "Every social relationship entails a state of indebtedness just as every state of indebtedness entails a social relationship" (Leach 1982, 159). Unequal relationships manifested themselves in unequal payments. They also showed themselves in how frequently accounts needed to be settled: higher-status occupations, as Leach noted, are compensated at longer intervals than lower-status ones (compare those with monthly salaries against those with day wages—and the difference holds well in East Africa as most anywhere). But hierarchy and inequality themselves can come and go, as Leach found in Burma, even in an oscillating pattern.

As anthropologists have noted in many parts of the world, wife givers are spoken of as superior to wife takers—or they behave as if they were, and they expect their recipients to follow suit, even if they still expect them to eventually transfer cows and other goods back. This is true of Luo, Nuer, and other Nilotic people, and of many of their non-Nilotic neighbors in East Africa (Evans-Pritchard 1951; Glickman 1971, 1974; Hutchinson 1996). Among farming Luo, indeed, it goes well beyond questions of marriage. Plow team owners, as we shall see, are deemed superior to plow team borrowers, and they make them work for them on terms that seem and sound even-steven on the surface, but are not underneath.

But it is not universally true that borrowers are, or feel, socially superior to lenders. The blade can cut both ways. Wealthy people who accept cash savings deposits for poorer ones for safe keeping, or persons who accept to foster others' children or animals to lighten those people's burdens of care— both common across Africa—are doing a favor that puts them in a way on top (even at the risk of taking on a headache) and gives them advantage in future dealings. This leads to the issue of class.

Power's Place in Fiduciary Culture

The political turn that economic anthropologists took in the 1960s and '70s, before the structuralist wave had passed, tied their work into other disciplines engaged in criticizing international capitalism. The scope of the discipline had broadened—that is, from the micro to include the macro scale. Dependency theories on metropolitan and elite domination, inspired by Marx and Engels but arising this time more from Latin America, reached Africa too, including Kenya (Leys 1975), spread partly through the influence of Walter Rodney's transcontinental studies. Usurious credit and debt, according to *dependistas*, as they became known, were among the devices by which European and North American centers actively underdeveloped Africa, much as they did Central and South America.

By the late 1970s and early '80s, Parisian and other economic anthropologists familiar with dependency theory work decided that capitalism did not just erode away other, precapitalist modes or relations of production (in Marx's terms), for instance, family, feudal, or plantation-slavery forms, but encapsulated and preserved them—"articulated" with them—for its own benefit (Godelier 1977, Seddon 1978, Meillassoux 1981). This way it could maintain a labor force without having to pay fair market price for it. Some "pre-capitalist" relations typical of rural Africa—for instance, marriage payments in cattle—were among these relations, by which elder men exerted control over junior men's and women's labor and reproductive power, and the order thus maintained may in turn be exploited by metropolitan capitalists or their rural agents (Meillassoux 1981, 67–74). Bridewealth payments were relations of credit and debt themselves, since transfers were often delayed. Big, metropolitan-centered credit systems might thus exploit more local, rural ones.

Meanwhile, in Paris, the highly influential philosophically and anthropologically inclined sociologists Pierre Bourdieu and Michel Foucault were by the early 1970s devising their own schemata of class and category exploitation. Bourdieu, in Kabyle in Algeria, observed that part of the power one gained by doing a favor for another was gaining control over the timing of getting one back: being able to command help when one might need it oneself, whether or not it was a convenient time for the one so indebted, was a big part of social strategy. Eliding sociology and economy, Bourdieu used the rather ambiguous term "social capital" for this kind of unwritten "account." For Bourdieu as for Mauss before him, ostensible generosity could conceal self-interest. Bourdieu also followed land pledging and observed the false "rescue" of indebted borrowers by kinsmen who ended up taking over their land, as one of the ways people used traditional culture as a mask to cover up self-interested or exploitative relationships.[12] He tended to observe individual thought and action constituting social structures and vice versa, cyclically and repetitively, in embodied cultural patterns ("structuring structures"). All this Bourdieu, again following Mauss and others, called "habitus" (1977). Bourdieu's land pledgers and their exploitation — manipulations on the Mediterranean — were not just a Parisian guru's perception. Legal anthropologist Sally Falk Moore (1986, 267–83 and passim) in the United States, placing more emphasis on active manipulations of fabricated "custom" and on historic change, noted comparable false rescues of land debtors by kin among Chagga in Tanzania: cold calculation on Kilimanjaro.

Michel Foucault's influence on the study of credit and debt in Africa was more indirect, yet still important. His readings on knowledge, power, and their overlap ("knowledge/power") influenced Africanist work on authorless, decentralized control of human thought. Our shared social institutions and discourse, by Foucault's way of thinking (and Bourdieu's too), affect the range and repertoire of thinkable thoughts within our mental reach. American anthropologist James Ferguson, a student of Sally Falk Moore who also followed Foucault, offered a close deconstructive analysis of a World Bank country study on Lesotho, and of the development industry for which the

Bank plays such a leading role. Against this report he set his own ethnographic knowledge of the eastern mountains of that small country and his reading of Lesotho history. He found the Bank and its officers to be creating and maintaining a set of spurious perceptions of aboriginal poverty that functioned to justify just the sort of technical and financial assistance (basically, credit and presumptuous oversight) that the Bank existed to provide. No coincidence here. The development apparatus was ideologically self-supporting. An added, "unintended" consequence was the entrenchment of a particular kind of state and bureaucratic power over rural people and their farming, labor migration, and life altogether (Ferguson 1990). Of course, rural people are not just dupes or pawns, and they have many ways to evade, resist, or subvert the idiom and the actions of lending and governing bureaucracies they resent. Or at least to try.[13] And borrowers who borrow big enough can gain a power of their own over their lenders. Borrowing and lending involve power, but it is not all one way. Nor is it always about force.

The Trust in Entrustment

The root of entrustment is trust, yet the concepts are not quite the same, and you can have one without the other. What, then, *is* trust? Who merits it for whom, and why? African people and anthropologists who have studied them have had something to say about this, contributing to conversations of philosophers and psychologists elsewhere.[14]

Whereas the *Oxford English Dictionary* (1971) gives as its first definition of trust "confidence in or reliance on some quality or attribute of a person or thing, or the truth of a statement," we might start by noting that humans need not trust each other just as individuals. One can trust a group or category, or a system or process in which people take part. People in rural East Africa, for instance, trust that when they sell their corn cheaply after harvest, others—it needn't matter yet who—will make available more later to buy, beg, or borrow by the next. (And the buyers, if speculating, trust the sellers or somebody else will be around then to buy some back.) The parties may be hard to specify. A Luo who borrows or lends a cow trusts others not to steal her and trusts the cow herself not to run away.

Trust is thought; entrustment is action with thought implied. If trust is in the head, heart, or (more to the Luo understanding of "spirit") liver, entrustment comes from these *and* from the hand too. Some moral philosophers including Karen Jones have suggested that, in her words, "we can entrust where we don't trust" (1996, 10; cf. Baier 1995). In this case, entrustment would be better labeled "loan" or "temporary conveyance" to avoid confusing action with sentiment. Thought and action may be only loosely related. Trust bridges a synapse between evidence and conclusion; entrustment bridges a gap between solitary inaction and social action.

To trust is to risk betrayal. What gives trust its value is the uncertainty about how someone or something will respond to an action or situation, together with the possibility that the response will disappoint. As sociologist Diego Gambetta puts it, "Trusting a person means believing that when offered the chance, he or she is not likely to behave in a way that is damaging to us, and trust will *typically* be relevant when at least one party is free to disappoint the other, free enough to avoid a risky relationship, *and* constrained enough to consider that relationship an attractive option" (1988, 219; his emphasis). But one may trust another not just to *act* (or not) in some way, but to *think* (or not) in some way.

Some, as seen already, equate trust with confidence or faith, but these terms themselves are not used quite the same way, as anthropologist Keith Hart (1998, 187) notes. Trust, as conventionally used, implies more doubt than reliance or confidence, yet more certainty than faith or hope. Trust connotes an emotional charge, hinting at a personal relationship, such as a friendship, having meaning and value in itself, or a spiritual one with a social side ("In God We Trust"). Trust is, as Karen Jones observes, an affective attitude: it can have as much to do with emotion as with logic or reason (1996). Nor, I agree with Hart, is trust the same as reliance. A stranded traveler must rely on a roadside mechanic he or she may little trust—while the latter might well trust, but not rely upon, this traveling stranger to return the spare tire or fuel can lent. In matters of government and public services in tropical African countries, including fiduciary ones, the difference between public reliance and public trust has lately grown wide.

Hart projects a continuum of English concepts onto a stylized schema of social evolution that he derives from nineteenth-century social thought. Referring to Henry Maine's *Ancient Law* (1861), he situates trust "in the no man's land between status and contract, the poles of primitive and modern society in evolutionary theory" (Hart 1988, 188). Trust, writes Hart, is "the negotiation of risk occasioned by the freedom of others, whom we know personally, to act against our interest in the relative absence of constraints imposed by kinship identity or legal contract." To these "constraints" one might add physical, political, economic, or military power.

Hart's case study describes a population of Frafra migrants in Accra in the 1960s as suspended midway between the kinship (Henry Maine's "status") of custom-bound, closed, traditional village life and the rational, open, civil society (Maine's "contract" of the modern world—see esp. his p. 192). Cut adrift from lineages, ancestors, and the strict norms of family and community, yet without the formal education or access to legal resources that would let them take advantage of national law, Frafra migrants turn, "almost *faute de mieux*," to free-floating association and friendship: "where self and other meet in some reciprocal understanding, where interests and risk are negotiated within relations formed by shared experience (even secrets), by love, knowledge, choice" (p. 189). No one in the Nima neighborhood lends to strangers, only to one's tenants or clients. The borrowers solicit loans by evoking friendship: "the pretension of familiarity is the normal rhetoric of economic life in a place like Accra" (p. 190). Hart accepts that at the urban extreme, real social relations are hybrids of kinship and friendship, obligatory and free association. The idiom of one tie is often used to reframe another.

In the end, the Frafras of Hart's study only partly succeed in making trust work: they feel their way along and many fail—perhaps after being stung. Trust is more brittle than kinship or contract: "Trust is central to social life when neither traditional certainties nor modern probabilities hold—in weak states or relatively lawless zones of public life, and in the transition to capitalism, especially in the mercantile sphere where credit is so important—but not as a basis for industrial production and division of labor" (p. 191). It's a useful halfway, but in Hart's Accra neighborhood it is not too sturdy a rung.[15]

By this way of thinking, people living in a country with worsening political upheaval and lawlessness—like the margins and resettlement areas of Kenya, or more extreme cases, Somalia or northern Uganda at the turn of the twenty-first century—would need to rely more and more heavily on trusting friendships, as well as on whatever kinship or clanship may perdure.[16]

But it is not necessary, in the present work, to adopt an evolutionary schema like Maine's or the weighty bag of assumptions about primitiveness and civilization that has tended for over a century to come with it. It would be misleading, I think, to try to peg Luo people or western Kenyans into a single point on the spectrum from status to contract. Real social relations are mixtures of both, in city (as Hart accepts) or country. Kinship and law can both impinge on friendship, clientage, and personal dealings of all sorts. And kinship does not disappear in towns and cities; it can just become more optative.

Trust may not fit the dealings of East African farming, herding, or fishing people better than does confidence or faith. By most definitions these can all be found there. Luo rural dwellers (and urban ones too) borrow and lend with brothers, with friends, and with complete strangers. They do so with trust, as when they send animals to a friend's for safe keeping; and with distrust, as when they enter into land mortgages that threaten to dislocate and dispossess them and their families for good. Sometimes they have little choice, as when sending cash home from afar with the only traveler going that direction.

Trust among kin needn't be warmer than trust between friends or even strangers. But the reciprocity it involves can certainly be warming. Psychologists offer broad comparative evidence (though it is mostly from North America and Europe) that, unsurprisingly, "those who are more willing to trust other people are likely to be equally trustworthy in that they are less likely to lie, cheat, or steal. They are also less likely to be unhappy or maladjusted, and are typically more liked by their friends and colleagues" (Good 1988, 32). So much Durkheim and Mauss would have agreed. The personal and affective dimensions matter, and the very fact of involvement in others' lives can outweigh calculable costs and benefits. In city or country, personal contact and the seasoning of familiarity — or lack of it — temper all exchange, all trust and cooperation. Even in lab games (Ostrom and Walker 2003).

Entrustment may create social ties, but ties exist only in the eye of the beholder, and an entrustment may not be equally welcome, warm, or enjoyable for truster and trustee. A secret, a stolen cow, or a fostered child can be a burden and a headache to take on, even if one is pleased and proud to do the favor. Even more unwanted may be the expectation that one ought to reciprocate with a trust or entrustment of one's own.

Trust and entrustment need not be simultaneous. You can lend for a lifetime to someone you've trusted only for a moment, and it is usually harder to *un*lend (that is, to call in a loan prematurely) than not to lend in the first place. Conversely, a brief loan, promptly repaid, can give rise to a feeling of trust that long outlasts it. There are other incongruities and asymmetries involved too. Distrust is *not* just a mirror image of trust. It can take a long time, and many repayments, to build a solid reputation for trustworthiness; but one false move, even a rumor, can blow it in a blink.

Trust is just a word used for a range of feelings, not always easy to translate, to gauge in temperature, or to place on an evolutionary scale. The trust behind an entrustment may be illusory, if there is act without sentiment. And yet the act itself can change the sentiment of the parties toward each other, or toward whatever moral community to which they may belong. Attitudes follow behavior more surely than behavior follows attitudes. Trusting does not necessarily lead to acts of entrustment, but entrusting is likely to make us more trusting.

Exchanges That Alter over Time

In exchanges that occur over time, the meanings of reciprocity change and get reinterpreted, and gift, loan, and theft or seizure can blend into each other in ways that arithmetic can scarcely reflect. Certainly the passage of time between prestation and counter-prestation can cement a social bond; in probably most cultures it is an insult to cancel out a gift with an immediate, equivalent payment or counter-gift. (In this subtle sense, "balanced" reciprocity can in fact be *negative* reciprocity underneath. Strict equivalency chills.) But time can also warp a social bond while cementing it. At the simplest, the lapse of time between loan and repayment allows much flexibility, but also potential misunderstanding, about the terms of the loan in the first place. Supplies, demands, prices, and values all shift in the interim. Where a loan is repaid in some form other than the one lent—grain for money, for instance, or labor for land—*both* things can change in value in the interim, thus obfuscating or rendering contestable their equivalency. A loan repaid in a different form makes interest charges easier to conceal—a crucial concern in some Muslim settings where "interest" is forbidden by sacred law.[17] Then there is the unpredictable. Currencies may get devalued, markets collapse, or governments fall; and a loan that seems fair at the time of a loan may no longer by the time of repayment. However ironclad the initial terms of agreement might be, debts mutate.

By cognitive slippage or linguistic sleight of hand, gifts may get reclassed as loans, loans as sales, sales as gifts. A land patron and client, in a hosting arrangement easy for all in times of sparse settlement, may later define themselves as antagonists when more settlers have filled up the land; and their respective children may disagree on whether the original invitation was meant to be a land loan or land grant. Individual acts get reconstrued as group acts, and vice versa. Loans can accrue obligation or be forgotten. Time can whet competition but may dampen it instead. These are matters of historical contingency, and of disputable and renegotiable agreements, not just of timeless rules and temporary infractions. This does not mean there are no patterns in the tricks memory plays—or the tricks people play with memory—but these are just beginning to be understood.

Whether most gifting or lending can be called customary or traditional, and just what that might mean, is debatable. Most social scientists up to the 1960s cast agrarian East Africans and others in the tropics as changeless traditionalists or cultural reactionaries needing to be modernized, but from the 1970s onward, as already seen, local "custom" began to appear more often as a manipulable and constantly reinvented tool—useful for accumulating land and capital, among other things. No longer could custom or tradition be assumed just an atavistic straightjacket for rural people, nobly helped by more enlightened foreigners, to escape.[18] In settlement and land use, marriage, child care, or funerals, people invent new uses for traditions, and they invent and reformulate customs.

Nor do locally organized modes of exchange necessarily yield to long-distance ways, or simple ones to complex, over time. Anthropological research brought the rotating saving and credit association, or self-help contribution club, into the limelight in the world of financial development, as a pervasive form of grassroots organization that provides financial services for people, most often women, who either lack access to banks or choose not to use them (Geertz 1962, Ardener 1964, Ardener and Burman 1996). This kind of group can be a simple, effective way of saving or borrowing while also helping others do so. The idea may spread from community to community, continent to continent, but the club needs the help of no superstructure, no bureaucracy. It demonstrates not a preference for liquidity, as some economists have assumed of humans anywhere—but instead a preference for illiquidity, or more precisely, for having wealth safe from the daily predations of oneself and one's kith and kin but still accessible in emergencies. Requiring no collateral for loans, but relying instead on peer group pressure, it challenges old assumptions about creditworthiness. And, since its finances derive from its members' own earnings, it suggests that more credit is not necessarily better credit. Some of these lessons have been gradually absorbed by an admiring development aid industry.

Along other, more theoretical lines, economic anthropologists made it clear in the 1980s and '90s that some conventional units of aggregation in economic research, particularly the individual, household, community, and even (and especially) nation, needed rethinking for African contexts, where they didn't always fit.[19] This and much other culturally self-critical scholarship contributed to old debates about whether economic life can properly be understood as beginning and ending with the individual (Long 1977, cf. LeClair and Schneider 1968), and these debates inform our analysis of finance, land tenure, and the links between them. The individual can hardly serve us well as sole focal point of analysis in settings where, for instance, fathers assume they can readily appropriate the cash earnings of sons, and where decisions about allocation of cattle turn as much on pressures to provide bridewealth animals for agnates as they do on considerations of personal profit. "Households" and "farms" turn out to be quite inadequate units of analysis for studying Luo and other settings where plural wives occupy multiple houses within a homestead, and where men and women keep largely separate finances.

In short, the cultural study of economic life in East Africa has challenged some of the deepest assumptions of European and American neoclassical economics and of the financial institutions that rely on them. Anthropology has raised serious doubts about whether societies progress from barter to cash to credit, about whether economy can or should ever be wholly distinct from other spheres of action, and about whether everything inexorably becomes exchangeable for everything else. The individual rational calculator now seems not always to be as rational, as exclusively self-interested, or as alone in real life as in economic texts up to the turn of the twenty-first century; and the units into which individuals aggregate differ widely from one society to another. Economic culture is political culture, full of power and privilege, and it is symbolic, moral, and aesthetic too. Goods and services, bads and disservices, and the time lapse in their transfer do not occupy their own world but are human contrivances, and they partake of the deeper, shared mental propensities by which humans structure a messy world. Fiduciary thought and practice connect time, space, and social distance in cultural ways not yet widely acknowledged.

CHAPTER 3

Luo and Their Livelihood

The Great Lake Basin and Beyond

The individual is synonymous with the stranger, an alien, possibly even an enemy. . . . You do not in an important sense exist until you reveal your networks.
—DAVID WILLIAM COHEN AND E. S. ATIENO ODHIAMBO, ON KENYA LUO

There are ways the Luo people could be called a microcosm of middle Africa, with its contradictions and subtle compromises. Most live inland, landlocked, like most of that great continent; but with the lake down the hill, and a city connected by road, rail, and an airport, some do have ways to come and go. Luo country, like Africa, defies generalization on ethnic purity or mixture. Situated in a part of the continent where they and other ethnolinguistic groups tend to live in rather solid, separate blocks on the map, the lakeside homeland of present-day Luo does yet contain a multicultural city they helped build, and rural localities of rich intermixture including Luhya, Luo-speaking Suba, and others.

Luo people take pride in their past, and are widely reputed for traditionalism; but they are far from stuck in it, and if you question a Luo who calls this people a herding or fishing people you are likely to find she has spent part of her life as an urban brewery worker (say) or nanny, and more still as a farmer. Call it hoe farming and you will be both right and wrong. Hoes with short handles are what women continue to use, and women do most of the actual work — yet there are ox plows everywhere (some driven, in a pinch, by women whose menfolk are away), and the occasional tractor. The Luo population as a whole can be called poor in material wealth by most African or world standards, yet not the poorest; in most of the numerous famines and plagues that have beset the lake region over the past century or two, most have survived.

Luo represent a Nilotic language, distinctively African, but among them there are many who code-switch to English and Swahili (with its heavy Arabic influence) when it suits them, and a few have learned one or another local tongue of their Bantu-speaking neighbors. Luo nearly all identify themselves as Christian, but scratch the surface of one or another Christianity and you will find a rich mix of local elements in it (black madonnas and saviors, witch and ghost exorcisms, blood sacrifice in crises, war dancing at funerals . . .) and in a few parts of the Luo country, strong Islamic ones too. You will also find more Christianities than you can count, for Luo churches divide and sub-divide endlessly as new prophets emerge and overturn old ecclesiastic orders.

Politically, Luo exemplify the kind of society that used to be called "state-less" or "acephalous" by a generation of anthropologists not too troubled by the negative ring of those terms; but one who looks into it deeper will find demi-chiefs and important diviners who served as war leaders. Even today a typical homestead has a special wooden stool, in effect a kind of mini-throne for that little realm, and shared legends, if not outright rules, that make it clear who may sit on it. There may also be a sacred spear with rules about who may take it outside the homestead. In a sense, then, Luo have monar-chies galore, just as they have churches galore. Their way is not so much anarchy as a diffused polyarchy.

Luo are sometimes contrasted with all their neighbors (Bantu-speaking and otherwise) as the ones without age sets, age grades, or circumcision ritu-als. Their detractors like to call them perennial children, since almost none, male or female, is ever subjected to genital cutting. But such banter is easily countered. Not only have many elders had six lower incisors removed in their youth—in personalized and private, not public ceremonies, yet ritual-ized and symbolically loaded ones—but school exams and graduations, now a central part of Luo cultural identity as a "tribe" of scholastic high achiev-ers, have rendered tooth extraction (*nak*), the nearest standard Luo counter-part to group circumcisions, all but superfluous. But Luo need not rely on schools for their maturity or dignity. Children are entrusted with the care of younger children; they do nursing and child care tasks that in other parts of the world would be legally if not morally restricted to adults, and thus grow up fast. Nor is there much equality later in life. Luo have their own ways, old and new, of sorting out first comers from late, seniors from juniors; and in Luoland one ignores these at one's peril.

The sexes, in Luoland, are superficially about as uneven in status as one can find in the world—in this too, Luo country is very African. But by now (shouldn't one expect it?) it has become clear that women, too, have power and in Luoland, not least merely by virtue of their eventual aging and repro-ducing. (It is sometimes said of Luo that "a powerful old woman" is a redun-

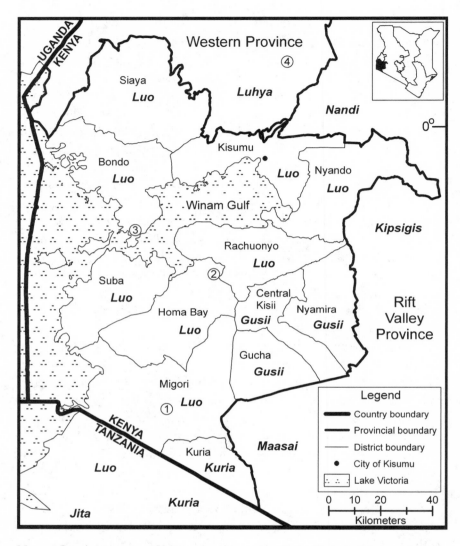

Nyanza Province, western Kenya, showing eastern Lake Victoria/Nyanza and predominant languages (in italics). Study sites: (1) Kanyamkago, main site; (2) Kagan; (3) Uyoma; (4) Isukha. Sources: adapted from CGIAR/IPC (2006) and World Bank (2006).

dancy, since age *is* power.) By outliving men, among other ways, women who lack power or influence in temporal or public spheres show that they may yet find it in sacred or private ones. And some of these avenues (divination, witchcraft, or cursing, for instance) seem *not* to be disappearing as people in the eastern lake basin learn new and foreign ways of thinking.

In short, it would be hard to find a more African place than Luoland, or a more African people than the Luo. The ways Luo handle foreign novelties are, as much as anything, what makes them so. If Luoland is a backwater, as it has often been described, it is a backwater with a city that at times has enjoyed steamer service to foreign cities; and Luo who turn the other direction and travel the rail "inland" toward the coast, on tracks Luo a century ago helped spike down, find enough Luo in the other cities to feel somehow at home.

Up from a Giant's Lake

Luo call a lake *nam,* and the one shown here Nam Lolwe. In one of their legends the giant Lolwe left the lake as his largest pool of pee. Other Kenyans call it Nyanza or still use Lake Victoria, its three surrounding nations having been unable to agree on whose current president to rename it after. The lake water, whatever its source, is fresh.

Moving inland from the lake, as Luo ancestors migrating to the region did, one starts out in grassy, lightly wooded savanna country but soon finds oneself in hilly land of quite different color and character, with more trees and bushes. Its topography varies from level ground, mainly around the lake shore (altitude 1,133 meters), to broad ridges and dissected verdant hills in the highlands. The arable parts of Luoland rise to about 1,500 meters at its highest boundaries to north and southwest, where it merges into Luhya- and Gusii-speaking areas. The Gusii highlands rise nearby to over 1,990 meters. Near Kisumu city, the sharply rising Nandi Escarpment has in recent times been a stark ethnic as well as physical divider. It contrasts sharply itself with the Yala swampland farther west, toward the strip of Samia Luhya country lining the Uganda border.

Luoland survives by rain, as there is little irrigation in the Luo country, the main exception being a government rice scheme around Ahero, in the flatlands southeast of Kisumu. In western Kenya, rainfall varies directly with altitude; the areas by the lake tend ironically to be much the driest and to have the least reliable rainfall, turning parched tans and browns each dry season. Because of this chancy environment, lowland farmers plant fewer and smaller crops than they know they could grow in a good year. But even in the lowlands, it can rain in torrents — as traveler Felix Oswald put it in 1915, here a thunderstorm is no shower but the bursting of a reservoir — and flood on baked hardpan soils, washing rather uselessly into the lake. Hailstorms too are a risk to farming in all parts of Luoland.

Rainfall in Nyanza, the province where Luo predominate, is bimodal.

The long rains (in Luo, *chiri*) usually last from about February to May, the short rains (*opon*) from about October to December. Both tend to be shorter in the lowlands than in the verdant highlands. This further limits what can be grown there. Not surprisingly, population densities in the lake basin tend to vary directly with rainfall.[1]

Soils in Nyanza vary enormously in type and quality, and farming people use sight, touch, smell, and taste to know these well. Dark red to red loamy sands are common in the higher lands, ash and pumice soils in the middle altitudes, and alluvial lithosols—the well-known, heavy East African "black cotton soils"—around the lake shore (Morgan and Moss 1972, soil map). Commonly two sides of a small hill or valley in Nyanza have entirely different soil types, suitable for entirely different crops.[2]

The Luo people have a widely acknowledged home territory in Kenya, where their ancestors have lived for at least several centuries. But they are not confined to it (they want not to be, despite a history of colonial confinement), nor are they its only denizens. Much the same can be said for their way of life and their ethnic identity. Surprisingly, for such a large and important group, no detailed general ethnography of Luoland has been published, but enough fine studies of Luo have been written that only the briefest introductory sketch is needed here.[3]

Of the 2,653,932 people classed as Luo in Kenya's 1989 government census—the last with a published ethnic breakdown—about 2,030,278, or 77 percent, lived in Nyanza Province, curving around the eastern shore of the lake called Victoria Nyanza or simply Nyanza.[4] They composed about 12 percent of the recorded Kenyan population of 21,443,636. Nearly all within lived in the three districts then called Siaya, Kisumu, and South Nyanza, from the southern part of which they extended into the northern Mara region of Tanzania, where significant numbers of Luo speakers, smaller and not lately differentiated in censuses but still many thousands, lived in an area adjacent the lake shore. The only district then in the Nyanza Province of Kenya not predominantly Luo was Kisii, enumerated as 98 percent Gusii. By the next census, in 1999, districts had been subdivided and renamed. Within Nyanza Province, the new districts in which Luo speakers predominated included (from the former Siaya District) Siaya and Bondo, (from the former Kisumu District) Kisumu and Nyando, and (from the former South Nyanza) Homa Bay, Rachuonyo, Migori, and Suba. Kuria (also part of the former South Nyanza) became a district in itself. The former Kisii District, home of the Gusii people, was split into Nyamira, Central Gusii, and Gucha Districts.

Some western Kenyans deemed the subdividing of districts a government favor (as an award of new districts), others looked upon it more skeptically as a strategy of divide-and-rule, still others as a simple expediency to accommodate a growing population. Accuracy of official censuses may be open to question, but all can agree that Luo heavily dominate Nyanza Province and the Kenyan third of the lake basin.

Linguists reclassify language families periodically, but at present one standard taxonomy files the Luo tongue (Dholuo or DhoLuo, literally the Luo "mouth") as follows. It is a subset of Luo-Acholi (distinguished from Alur-Acholi), within Southern Luo (which also includes Lang'o, Labwor, and Adhola), within Western Nilotic, within East Sudanic, within Nilo-Saharan.[5]

Evidently Luo are not northern anything. And that is how they see it too. Common oral traditions of the Kenya Luo suggest that their ancestors migrated south-southeast in small waves, in stages, and by more than one route from the Bahr el Ghazal region of the Sudan, around the homeland of the present-day Nuer and Dinka, the first of them entering the Nyanza region perhaps some 450 to 500 years ago or earlier.[6] They seem to have taken more than one route. Some think migrants forced or encouraged each other south, in Edward Evans-Pritchard's oft-repeated phrase, "like a line of shunting trucks." But there are other ways it could have happened. I think it likely some of the movement was more piecemeal, more a matter of families and individuals in hard circumstances joining kin or friends who had gone ahead and sent word back it was safe.

The divisions of migrant and settler groups are the stuff of mythicized history, and some of it follows recognizable patterns wherever Lwoo people live, from Sudan to Kenya and Tanzania, and even abroad. One of the most conventional moral tales, an origin story with many variants, is crucial for our topic, since it concerns property and obligation. It concerns a lost spear and a lost bead—and two events of reciprocal nonforgiveness, which some say caused a rift at a place now in northwestern Uganda called Pubung'u. Here is a western Kenyan version taken mainly from Onyango-Ogutu and Roscoe's rendition (1981, 133–8).

Ramogi (or Olum) has two married sons living in his home, Nyipir and Nyabong'o. One day Nyipir borrows a sacred spear (*sepe*) from his brother, Nyabong'o—or rather, takes it from Nyabong'o's house in the latter's absence to spear an elephant who has strayed into the homestead's crops. His throw hits the elephant, but then he loses the spear as the elephant runs away with it. Nyabong'o, upon returning, demands the sacred spear be returned, insisting no number of cows and no other spear could substitute for it. Nyipir sets out on a hard journey. On it he encounters the ugly, dirty, and almost blind elephant mother, Min Liech, with whom he shares his food provided by his wife. The elephant shows him to a collection of lost spears the elephants have left. He finds Nyabong'o's spear in that collection. Happy at Nyipir's good fortune, Min Liech gives him a gift of beads. Nyipir returns home and gives the spear back to Nyabong'o. Nyipir gives the beads to his own wife with a warning to guard them. She sits stringing the beads. But Nyabong'o's crawling child eats a bead before anyone can stop its happening. Now it is Nyipir's turn to forgive and forget or not. He is furious and demands the bead be returned. No attempt to find the bead in the child's waste succeeds. After three days, the child is cut open and killed, and eventually the bead is found. A violent quarrel arises between the two brothers' families, splitting the house of Ramogi. Nyipir and his followers migrate west to Alurland, west of the Nile, and Nyabong'o and his go south to Acholiland (some to continue beyond). "And that is why the Luo migrated south to Kenya, and even as far as Tanzania."[7]

This moralized tale, blaming an eventual social rift on a mishap and on Shylockean avarice, instructs young Luo not to be too concerned with getting their own back as individuals, but instead to share generously with kin — and also with strangers in need like the ugly old mother elephant — and to forgive. Intimates can be rivals too. Life is most precious, and possessiveness divides. The pursuit of strict reciprocity is not worth risking peace and unity. This is not just history; it is training for borrowing and lending, and for life.

All evidence on the early chronology remains highly speculative, but there is much evidence that ancestors of present-day Luo first moved into the lake basin along the lake shore itself, and that they displaced (ancestors of) western Maasai and perhaps other groups by gradual conquest and encroachment, fanning eastward up into the hilly areas that now constitute the most fertile and densely settled parts of their lake basin homeland.

That Nilotic speakers became so numerous in what is now western Kenya reflects the absorption of many Bantu-speaking people (who may have originated from southern or central Africa) and possibly others into their society as they moved southward around the lake basin and then uphill. It appears that many of these people entrusted their well-being to Luo immigrants and conquerors who allowed them to survive peaceably, perhaps under the condition that they adopt Luo identity and customs. These conquered or otherwise subjugated people blended in partly through intermarriage and fictive kinship. Luo migrations probably slowed down by the time Kenya was claimed a British protectorate in 1895, but they did not stop. Local and interethnic warfare too may have slowed down in what Anglophiles have liked to call the Pax Britannica, but that did not stop for good either.

Population densities of the Luo country are now very high, by African standards. Nyanza Province averaged 350 per square kilometer (as against a national average of forty-nine) in the 1999 census, most of the Luo districts averaging in the two and three hundreds (vol. 1, p. xxxiii). (Still higher, indeed some of the highest in Africa, are densities in the Gusii-speaking districts of Nyanza and Luhya-speaking districts of adjacent Western Province — these rose to 700 and above.) Densities remain high despite the ravages of rinderpest and pleuropneumonia at the end of the nineteenth century, and of HIV/AIDS, malaria, diarrheal disease, and numerous other health scourges in recent decades. For HIV and malaria, western Kenya has among Africa's highest (though still not precisely known) rates of infection, the former epidemic indeed having wiped out a large part of the middle generation in Kenya. Kenya has enjoyed, and also suffered from, one of the world's highest birthrates, measured at about 3.4 percent per year in the ten-yearly censuses of 1979 and 1989 but subsiding to about 2.9 percent by the 1999 census (vol. 1, p. xxx). Progeny give a man and especially a woman status and prestige — the more the better — and heirs to keep one's memory alive. A history of territorial confinement under colonial rule, and the expense and visa difficulties today of travel abroad, have also kept local densities high.

Poverty and disease, often appearing arm in arm, arguably contribute in turn to high birthrates. It is on the young that the old must rely when too old to feed, clothe, and house themselves; and lately chances to find well-compensated work, at home or abroad if one can, have been scarce enough to make many accept multiple births to beat the odds. Vulnerability also makes it an important and adaptive survival strategy to give, lend, and otherwise entrust enough to neighbors (but not to expend too much on non-reciprocators or wastrels) to guarantee sources of help in times of hunger or illness. Population density, whatever Malthusian hardships it may foretell, is not all bad where no one can survive alone. Most Luo, like other people, try hard to ensure they have some sort of family, friendship, and community no matter what. In East African eyes, as a general rule, solitude is bad and company good.

Settlement and Society

Between what people in East Africa say they do for a living, and what they really do, often lies a world of difference. Most rural-dwelling Luo, while deriving their main livelihood from cultivation supplemented by intermittent wage work and remittances, like most others in the region—like to think of themselves more as herders, fishermen, or white-collar workers or aspirants, as though to remind themselves to cut risks by keeping their portfolio diversified.

The Nilotic linguistic roots of Luo have led many to suppose that they once lived a more migratory life. But there is little hard evidence that the Luo were ever a nomadic or even transhumant people, and it is clear that at least since the great rinderpest epidemic of the 1890s, they have lived mainly by cultivation.[8] Over the past century a shift toward heavier reliance on agriculture at the expense of herding has been a practical necessity, as colonial and international borders have curtailed Luo migrations and expansions, and as the Luo population has grown increasingly dense.

The settlement patterns of the Luo people have changed much since the late nineteenth century and decades before, a period when many lived in stone- or earth-stockaded defensive settlements called *gundni bur* (sing. *gunda bur*), each with numerous houses and granaries within, and with fields surrounding outside. The remains of these can still be found dotted around Nyanza. While some historians deem the Pax Britannica the crucial historical point for the spreading out into separate homesteads (*mier,* sing. *dala*) for monogamous and polygynous families, others contend that such a change had begun long before and was more gradual.[9] And while some have supposed membership in these stockaded settlements to have been based on patrilineal kinship, like much of Luo rural settlement in recent decades, evidence on seven *gundni bur* in Siaya suggests group ascription and recruitment not so much on the basis of patriliny, but marital ties and non-kin connections (Cohen and Atieno Odhiambo 1989, 13). This fits other evidence, in the Luo country and elsewhere, that segmentary (branching) patrilineages as obtain today in Luo behavior or ideals are *not* merely an atavistic residue of a primordial East African way of life, but instead more of a modern adaptation to competition for land and other resources.[10] Any trust and cooperation discernable among agnatic kin at present, accordingly, are better explained in terms of present conditions, and of contemporary purposes or functions.

Like their Bantu- and Kalenjin-speaking neighbors, Luo people in rural areas live now for the most part in scattered homesteads rather than nucleated villages. In the 1999 national census, Nyanza homesteads averaged 4.5 members.[11] English terminology for family and domestic groups is problematic in western Kenya, as in much of eastern Africa. In this study a "homestead" refers to a single or multi-house compound (dala) of the sort usually based on an elementary, nuclear, joint, or extended family. Plural wives usually occupy their own houses (sing. *ot*, pl. *odi* or *udi*) within a homestead and control their own fields and granaries and other food supplies, together with their respective resident offspring.

The social organization and culture of the Luo are familiar to Africanist and other social scientists through books and articles by Edward Evans-Pritchard, Aidan Southall, David Parkin, B. A. Ogot, David Cohen, E. S. Atieno Odhiambo, Margaret Jean Hay, and a number of others. Some main findings from more than a half century of writings snowballing around a few sources can be pretty simply summed up for the moment.

Among the political entities deemed traditional, the largest, the *oganda* (pl. *ogendni* or *ogendini*) or federation of contiguous patri-clans (but sometimes called a tribe or subtribe), was much smaller than the Luo people as a whole, reaching into the tens of thousands in the early twentieth century when colonial government formed its first administrative "locations" on this basis (see Evans-Pritchard 1965a). In its territorial aspect, an oganda is called a *piny* (pl. *pinje*), country. Typically it had a leader of influence (*ruoth*, pl. *ruodhi*) whose authority varied and could only in a few cases be strong enough to call him a chief, and a man of medicine (*jabilo*) who usually doubled as war counselor and sometimes was deemed a rainmaker.

The Luo have become a textbook case of a formerly stateless, segmentary (that is, branching) lineage society. Patrilineal in descent group recruitment, patronymic, virilocal in postmarital residence, and, in a large minority of families, polygynous, the Luo put males in the public limelight, often concealing substantial female influence and discussion between the sexes in decision-making.[12] While Luo do not go in for age sets and age grading in the way that some neighboring East African herding and farming people do, matters of seniority are no less critical. In the past the society has tended strongly toward gerontocracy in familial matters—at least in their overt expressions—and many public ones.

But the standard characterizations about patriliny among Luo (and about other East African people around them) also need some qualifiers and disclaimers. One is that although Luo do seem to carry in their heads some models linking branching agnatic kin groups to spatial order on the map (or, to borrow a concept of Paul Bohannan's, a stretchable map to grow with), and many do indeed settle where their genealogies would place them on those maps, the exceptions are at least nowadays about as common as the rule. Some women, upon marrying, do not simply move to their husbands' homes and stay there with them to reproduce there as the standard model would have it, but instead move there only temporarily with them or stay put in their own homes and bring the husbands there (uxorilocality, the reverse of virilocality) to enjoy circumstantial benefits like more land or perhaps nicer neighbors. Or they find land to settle on away from both (neolocality). Where a man has requested land away from his paternal home and moved there, his family begins to form a new patrilineage there in its new home. For two or three generations at least, they and their progeny who stay around are treated as something hard to translate into English: *jodak* (sing. *jadak*), that is strangers, transient dwellers, land clients, or allochthons (that is, non-belongers); among *weg lowo* (sing. *wuon lowo*), literally "fathers or masters of the land," or natives or autochthons (that is, people who belong).

The second point is related. Some of the earlier findings about the dominance of patrilineal principles have been challenged by a wave of revisionist thinking that accords more importance to matrilateral and affinal kin ties (see, for example, Cohen and Atieno Odhiambo 1989).[13]

Finally, the role of friendship, as distinct from kinship (though often overlapping with it), has been seriously undertreated in studies of Luo and all Nilotic and East African people. Real human actors play more than their scripted roles in social structural models. They also like to make up their own minds about whom to admire and associate with. Some of this association they put into action, and bring into reality, by what they entrust the people they pick.

Luo landholding is about trust: both trust in past and future generations, and trust—or distrust—in contemporary allocative process. Until the British colonial government began trying to transform Kenyan landholding into a system of private, individual property in the last decade before Independence in 1963, Luo held land by a complex though gradually changing system of allocation and redistribution that depended upon several principles simultaneously. Because it has been described in several other places, only the most important features need be summed up here.[14] But these features are crucial for understanding the outcome of Kenya's rural credit system, among other things, as they directly affect whether a mortgage system can be created as national development planners have hoped. In their landholding pattern Luo have quite a bit in common with the Gusii and southern Luhya peoples who neighbor them, and some similarities with other large western Kenyan groups.

Into and throughout the colonial era, and to a large extent still, the key principles were that one could gain and maintain access to arable land only *by membership* in a kin group and broader community, with all the good behavior that that involved; *by labor* on it; or both. Rights of individuals were not thought sacrosanct, but instead they interlocked with the rights of others, and overlapped with those of families and wider groups. Rights of access and use oscillated seasonally between individual and family cultivation rights on a piece of land during a growing season, and more open grazing rights for a broader community on the same land after the harvest. For most land, access for passage and for collecting spring water, wood, thatch, and other materials was normally open to neighbors. And rights entailed duties: holding arable land meant guarding it in custody for members of past and future generations.

A place on the landscape implied, and still does, a place in a kin group, and vice versa. Patriliny, virilocal residence (settling at the husband's natal home after marriage), and the subdivision of holdings devolving from one generation to the next have been the norms in the Luo country since at least the early decades of the twentieth century. After marriage and the birth of one or two children, a Luo man and his conjugal family most commonly set up a new homestead adjacent to or very near the man's father's. The ideal pattern is long established: the first son's homestead is placed near the father's doorway or gateway and on its right side looking out, the second nearby on the left, the third just beyond the first, the fourth just beyond the second, and so on. The descendants are expected to push outward from the paternal homestead, setting up new homesteads and subdividing lands in thin parallel strips extending outward, as the family multiplies. Where enough land is available, the homesteads' layout can provide a physical map of a family tree.

The segmentary clans and patrilineages by which Luo are known are designated by several terms.[15] These include *jokakwaro* (descendants of a common grandfather), *dhoot* (doorway, literally "mouth of the house"; figuratively referring to descendants of one woman), *anyuola* (from *nyuolo*, to give birth), and *libamba* (a large and genealogically deep lineage group), according to size, depth, and conversational context.[16] As noted earlier, here as in some other parts of Africa lineages have lately been debated among scholars as being more the idealized products of indigenous or exogenous imaginations—and male ones at that—than a reflection of real settlement patterns and landholding layouts. But in the Luo country they are social facts of great importance. At least since the early twentieth century, the dead have been buried in their family homesteads, and cultivation rights and broader claims of belonging have been justified by reference to ancestral graves. Men have acquired access to arable land mainly by clearing or inheritance inter vivos but also by gift, loan, swap, or gradual encroachment on neighbors' fields. Elders adjudicating disputes have sometimes tacitly condoned encroachments if these seemed to even out some maldistribution.

Women have not usually inherited cultivation rights in recorded history but have acquired them mainly through marriage. As foreigners have usually understood the picture, women's rights are only ancillary, depending on allocations from their husbands. But since women do most of the farm work in Luoland, it is more useful to see the picture another way: women obtain rights in the areas of their postmarital homesteads by devolution from their mothers-in-law, who supervise them, or at least purport to do so, in their early married years. Women, like men, sometimes swap fields between homestead groups for temporary use.

Land exchanges break up somewhat the ideal settlement pattern in reality. A well-established tradition of long-term land loans and land clientage—all a form of entrustment, among other things—allows domestic groups to move between the territories of lineage segments, and it makes for the fragmentation of these segments over time. By this process, men have long gained cultivation rights for themselves and their wives as jodak by borrowing land in the home territories of their in-laws or strangers, and moving there, accepting subordinate social status by doing so. A possibility of expulsion or ostracism, nearer at hand for a land client than for a wuon lowo, land patron, served to keep personalities agreeable. But to bury one's dead kin on the land solidified rights in it. And many land clients, after starting out as borrowers, have stayed on the land for good. Indeed, descendants of land borrowers have sometimes even crowded out those of the lenders.

Far from being atavistic residues of a primordial condition, the patrilineal and virilocal principles appear to have grown gradually *more* pronounced in land matters over the colonial period and after Independence. The reasons are several. Rising population densities, cash cropping, and new labor-saving technology like the ox plow have all sharpened competition for arable land. Land shortage has made it harder for farmers to move their homes as they exhaust their soils or perhaps fall out with neighboring relatives. Easier, under the circumstances, is to stay home and rely on receiving parts of sub-divided paternal homesteads instead — or, from women's perspectives, to rely on land from mothers-in-law. (Similar principles and trends have been ob-served among the Luhya and Gusii peoples, in western Kenya; and some were discernible also among other densely settled groups in central Kenya and in some other eastern African countries.) But land clientage between in-laws and friends has still continued.

The complex but adapting Luo system of landholding has come under strong challenge from the colonial government and successive regimes since the time of Independence. The overlay of a system of freehold titling, a foreign-introduced program aimed at creating easily transferable and mort-gageable land, has by no means expunged the local landholding patterns there before, nor is there much reason why it should. The juxtaposition of two radically different systems of rights and duties is a source of major and continual cross-cultural misunderstandings in the Luo country, and in much of the rest of western Kenya.

What Luo Grow (and Whom)

Into the twentieth century, Luo people farmed with short-handled hoes. The ox plow, introduced by 1911, has been gradually taking over, though hoes are still used most everywhere. Roughly half the homestead families in Nyanza have their own ox plows but many more are able to borrow or hire them.[17] The main food crop grown and consumed in Nyanza, and the staple of most of Kenya, is maize (*Zea mays*) — that is, corn, originally from Mesoamerica. It can be grown at all altitudes of Luoland in good seasons, but little is planted in the low and dry lakeshore areas. Maize has largely supplanted sorghum (*Sorghum vulgare*) and the more labor-intensive finger millet (*Eleusine coracana*), the more drought-resistant former staples, in Luo-land this century; but these, and particularly sorghum, remain important in the drier lower altitudes. First domesticated in Mesoamerica but grown in parts of Kenya since the seventeenth century or earlier, maize gained popu-larity in the Nyanza region between 1900 and 1920, when European settlers introduced white varieties from South Africa to Kenya. Many upland Luo farmers today devote more land to maize (often intersown with beans) than to all other food crops combined. White maize, machine milled or hand ground, is now the preferred substance for the Luo *kuon* (Sw. *ugali*), the semisolid mass of boiled flour, the firm grits moldable in the hand, that provide the main bulk and calories in most Luo meals.[18] Without kuon, Luo say, no meal is a meal. Sorghum, millet, cassava (or manioc, mostly *Manihot esculenta*), or mixtures of these are used; Luo and other western Kenyans turn to these if maize runs out. Most deem them less-sophisticated foods, though a reverse snobbery for hearty peasant grains can be seen among some urbanites with a choice, here as abroad.

All of these and sugar cane may be used in various mixtures for local beer (*kong'o,* Sw. *pombe*), a traditionally important ceremonial drink, or porridge (*nyuka*), a filling snack, both significant sources of calories and protein. Cassava, which like maize originates in the Americas, has been used in the twentieth century as a famine relief crop. Its bulky and starchy though vitamin-deficient tubers lie in the ground year-round until needed, impervious to locusts and most other crop pests. For this the Luo nickname cassava *dero oko,* the granary outside the homestead.

A wide range of other vegetables rounds out the Luo diet. Most important are sweet potatoes (*Ipomoea batatas*), English potatoes (*Solanum tubersom*), and the many varieties of beans (*Phaseolus vulgaris*), the last of which are usually intercropped with maize. Miscellaneous other pulses grown include green and black grams (*Vigna aureus* and *Vigna mungo*), pigeon peas (*Cajanus cajan*), and cowpeas (*Vigna unguiculata*). Groundnuts (peanuts), grown mainly for sale, are an important cash source in the lower altitudes. Other crops include a large number of local leafy vegetables whose symbolic importance may exceed their economic (Schwartz 1989); and bananas, pineapples, mangos, pumpkins, citrus fruits, and sesame. Kitchen gardens are devoted largely to kales (called *sukumawiki,* a Swahili term meaning to push or stretch through the week; these are so called because they can be harvested at any time). European-introduced vegetables, including string beans, cabbages, tomatoes, carrots, and avocados, are grown, but not all these are widely consumed in Luo homes.[19]

All the food crops Luo grow may be bought and sold in local markets. Those most widely sold, and those sold in the largest quantities, are maize and beans. The latter and groundnuts, of all the main food crops sold, fetch the highest values per unit of weight or volume.

Crops grown almost exclusively for sale include cotton in the lower zones, coffee in the upper zones, new varieties of sugar cane, and tobacco. Luo farmers today grow sisal to mark the boundaries of their fields, and many sell the cuttings to provincial factories or braid rope at home. Cotton, introduced to Nyanza in about 1904, is the only major crop grown solely for cash in the lower zones of Luoland.[20] Of all Kenya's industrial crops, none has had a longer history of market inefficiency and production disincentives than cotton. The western provinces have seen eighty years of recurring price slumps and unfulfilled promises of infrastructural developments. At home or away, some Luo have also participated in growing flowers for immediate export on harvesting—like string beans, a recently booming horticultural development involving contract farming in the more fertile parts of Kenya. On many of their own crops Luo use animal manure; on some, synthetic chemical fertilizer. Indigenous means of bug and worm control—for instance, charcoal and plant deterrents—are supplemented by purchased or loaned chemical pesticides, used mainly on cotton and tobacco and sometimes on stored maize.

Luo people are quite capable of hard work in farming when the incentives are right, but generally these have not been strong enough to earn them a reputation as great farmers such as Gusii or Gikuyu have earned in their higher, more reliably watered altitudes, long favored by developers of transport and storage infrastructure. Colonial agricultural officers sometimes saw fit to co-opt chiefs and subchiefs to beat Luo people into growing cotton as a cash crop and cassava as a famine crop, which did little to enhance the farming zeal.

Luo and other western Kenyan people depend on several kinds of farm animals, but some animals matter more to their cultural self-image than others. Large livestock are a main preoccupation of Luo farmers, particularly men. Nilotic-speaking men, including Luo, are famous for identifying personally with their cattle, even when it is women and young people who stay around home to care for the animals and work with them in men's absence. Cattle play numerous crucial roles in culture as well as economy. They have been and are still used as standard ritual and exchange items in bridewealth (made over time from a bridegroom and his natal kin to a bride's natal kin), funerary distributions, and other symbolic slaughters and exchanges.[21] Luo also keep sheep, goats, and chickens; and in a minority of rural homes, donkeys are kept for haulage or traction. Owners diversify their herds and flocks to take advantage of different feeding needs, reproduction cycles, and disease vulnerabilities.[22] Cattle are sold to local butchers in times of urgent cash needs; otherwise they are slaughtered only on important ceremonial occasions such as weddings and funerals, on which occasions particular parts of the animals are assigned to particular kinds of kin.[23] Luo and other western Kenyans are likely to slaughter sheep, goats, and chickens at home only

for honored guests or for sacrificial offerings in times of special need or uncertainty. Luo have long traded livestock and grain (the direction of flow
depending) with people of neighboring Bantu-speaking and Maasai groups,
and much of this exchange goes on outside of town marketplaces.

Like many other peoples practicing mixed farming and herding in Africa
south of the Sahara, Luo have had in the past a hierarchy of values whereby
men seek to convert grain to small stock, small stock to large stock, and large
stock to wives (and thus also to children) when their circumstances permit.
"Downward" conversions of stock to grain are considered emergency measures.[24] Today all these exchanges may involve cash.

Most cattle are considered male or male-overseen property, and a form
of saving and investment for the future of the individual and lineage. They
are the center of a prestige system fundamental in the Luo social order: men
seek to consolidate their interests as against those of women, and elders to
consolidate theirs as against those of juniors, by converting wealth to cattle
and retaining it in this form. Older men thus try to perpetuate a one-way flow
of wealth from cash to cattle.[25] Converting wealth to cattle removes it from
easy negotiability within the homestead economy: women and children can
more easily make demands on male wealth (if they know it exists) when it is
in a more liquid form, like cash, than when it is in cattle.[26] But in Luoland the
rising population pressure on land is now threatening the Luo values on cattle
ownership, even among elders.[27] It also threatens the high value on polygyny.

Whether animals are like money is a subject of endless debate and misunderstanding in the Luo country as well as between anthropologists and
economists outside. In some ways they are and in others not. But it is clear
that there and in much of eastern Africa, they are the currency, or medium,
of lifelong entrustments and debts involved in marriage dues and their payment. Most Luo marriages and the payments from grooms' families to
brides' families occur gradually over many years, and they cannot be called
complete until the last cow is handed over—and in many cases it never is.
Protracted marriage dues place Luo under lifelong obligations to in-laws
who have provided their families and lineages with wives and give them lifelong claims over those whom they have in turn provided with wives. In Luo
eyes, the claims of distant creditors like cooperatives, state banks, or international aid agencies pale in comparison with the perpetual, living and breathing debts between in-laws; and this is among the reasons, for better or worse,
why institutional financial loans tend to be repaid as scantly as they are.[28]

Thirsty Water and Uncharted Ponds:
Fishing Culture and Aquaculture

Of all the many kinds of hunting and foraging Luo sometimes do, fishing has long been the most important, and as a people they are well known around Kenya and Lake Victoria/Nyanza for their interest and skill at it. They do it by more methods, and invest more meanings and emotions in it, than the following brief sketch can describe. They also make more kinds of gifts, sales, and loans of fish and fish products, and also indeed of the means for catching them. Much of what is said of their customs and adaptations here applies to other lakeside people (especially Samia Luhya in the northern shore, Suba in the southern and in Rusinga and Mfangano Islands). Luo call themselves *jokanyanam*, (the) people of the lakes and rivers—a term with evident allusion to their southerly migration along the Nile—or just *jonam*, people of the lake.

Luo people fish the big lake, a small satellite lake (Kanyaboli, north of the gulf) and impoundments, the rivers, and lately human-made ponds. They fish individually and in small and large groups. The lake fishing they do from shore and by sleek sailing canoes (usually planked, less often dugout) and other sailboats; by single-blade paddle, pole, and oar-propelled boats and rafts; and lately by motor launches and larger commercial boats, some of these last foreign-owned. Mostly they use the dragnet (*gogo* or *par*) and single or multi-hook line, sometimes suspended stationary by poles or floats. They team up in twos and threes, sometimes more, when in small boats (*yiedhi*, sing. *yie*)—some of which, like road taxi vans (*matatu*), are painted with bright and imaginative designs, in this case often abstract. The weir or enclosure (also *gogo*) of reeds sometimes woven in mats, or of poles; the basket cone trap (*aunga* or the smaller *nyadudi*); and the spear (*tong'*) or harpoon are among the other traditional methods used in the lake and rivers. Some of the devices are not made on shore and taken out onto the lake but the reverse, cut from floating reed beds and then landed ashore. In an occasional big lake shore fish run such as elders like to remember, all nearby including children might partake in a day's wading chase, coordinating in pairs or groups for scaring the fish into baskets and other traps.

Much as Luo people like to think of themselves as fishers, never in known times has fishing been something everyone did as a regular practice. Far from it. Boat building and lake fishing have remained specialist minority occupations for shore communities, and in those only in certain families' traditions. They are activities with much oral lore learned from childhood, and diverse taboos (*kwer*) on foods, sexual relations, and menstrual blood contact to help cope mentally and materially with real dangers of the lake. Its being land-locked makes this giant lake hardly less dangerous for fishers and small craft, given the hippos and occasional crocodiles at its edges, and the suddenness with which its empty, placid horizons can yield to sudden whirlwinds and lightning storms. Snail-borne disease, especially schistosomiasis, is a subtler lake shore danger.

New nets and boats, in the past and present, have been approved by di-viners and blessed and "cleansed" in elaborate ceremonies that can involve dozens of people and last several days. New boats are treated as wild and dangerous unless properly tamed or socialized by ritual. Luo people speak of boats in an idiom used for kinship, naming them sometimes after daughters or referring to them as brides for their owners. Launchings include several named phases (rwako, dwoko, riso) that appear also in people's own mar-riage ceremonies, including libations and sacrifices (described in Cohen and Atieno Odhiambo 1989, 99–103). Boats may be attributed personalities and intentions and pleaded with or cajoled, or in other contexts treated matter-of-factly like just so much wood and rope. Lakers who fish together in boats, which on the sleek, tippy ones means placing their lives in each others' hands, yet organize themselves in marked hierarchy, an order mystically sanctioned. Not only a boat owner but also a *net* owner claims special privileges in di-viding a catch. No less important in a Luo understanding, a boat owner (or if there is no boat, a net owner) claims the privileged place, usually first, in ritual sequences of many sorts (examples in Mboya 2001, ch. 6). The order extends to their spouses: the first catch to the boat or net owner's first wife, and so on. In the lake basin just as in Melanesia or the northeastern Pacific rim, and maybe anywhere, it is the more dangerous sorts of boating that in-volve the most elaborate ritual preparations and proscriptions. Lake Victo-ria/Nyanza takes surrogate sacrifices when it and ancestors associated with it are deemed, in Luo story writer Grace Ogot's words, "thirsty for human blood" (1984, 148). A goat might do for her troubled fisher, for an offering of appeasement and for making a hide and herb amulet (which he later sus-pects of causing more trouble—how now to get rid of it?). Many Luo and other lakers express worry if a new fishing boat is launched without a sacri-fice of a rooster or other fowl that I have heard the occasional observer—in humor, horror, or resignation—call "the chicken thing."

In fishing-related rites the ideal sequences, at least the ones deemed tra-ditional as inscribed in a lapidary text like Paul Mboya's, can be tortuous or tempting to break (the owner must sleep with the first wife first . . . . the first fish from the new spear must be eaten unsalted . . . . one should not look into the other's pot until . . . .). And that would seem to be part of their point: they are numerous and arcane enough as if made to be broken or half-fulfilled somehow and in this way always to leave a bit of uncertainty. This makes the broader conceptual order resistant to falsification. A flouted taboo of ceremonial sequencing, a neglected sacrifice, or a botched prayer conveniently helps explain the sorts of drowning or injury events that may otherwise be explicable only as chance or bad luck—a notion not all Luo or lakers accept—or as fate. Capsize drownings and injuries on the lake keep occurring, and these concerns of protection and explanation remain as serious and earnest as ever.

Boats nowadays require licenses, and thus official or unofficial fees, but there have always been fishers who go without. For Kenyans (whose nation claims only 6 percent of the lake's surface), Ugandan and Tanzanian patrolmen are among the navigational hazards. It would be hard to imagine their staying within those invisible lines, and indeed they do not; nor do Tanzanian and Ugandan fishers keep out of Kenyan waters. It is hard to say what role personal contacts and unseen reciprocities (and prayer and magic) play in the chasing, hiding, and negotiating, or what unofficial diplomatic dealings might occur between the fisheries and customs authorities from the three nations. What is sure is that the great lake's boat fishers count on exchange and often entrustment, whether for fuel, free movement, or protective blessings.

Lake fishing by boat has typically been a male activity, but when fish reach the rural shore women take over much of the smaller-scale processing, transport, and marketing. They and men buyers do some of their purchasing on short-term credit from accustomed boat owners they know. Drying some of the fish on the shore or taking it whole, they do much of the transporting inland, by any means including women's head load, to resell in marketplaces, in their own or through other traders' stalls. In fish as in farmed crops, though, men have hitherto dominated the larger-scale wholesaling and longer-distance marketing, so often leaving women and children to process, scavenge, or sell the small fry and discarded fish parts. But that is not quite the whole picture, as Karen Flynn points out (2005, 45). She describes another sort of women in the marketplaces on the southern, Tanzanian side of the lake shore: "richly clad Zairean buyers wearing brightly colored dresses, lots of gold jewelry, and turbans. They made the long drive to Mwanza to sell Zairean gold and filled their massive trucks with dried *dagaa* (tiny sardines [or minnows]) and other fish to sell back home."

How did those fish get so small? Whether one construes it as a story about humans or finned creatures it is indeed a story about big fish as well as small, and also about both locals and strangers. It would be easy to pin-point one villain or another (and not just human, but animal, vegetable, or mineral), depending on when one began the story, but the story's implications are wider and deeper than that.

In brief, local fishing slowed considerably over the past half century, and especially between the 1970s and the turn of the century, as a result of dras-tic ecological and economic shifts in and around Lake Victoria/Nyanza that many have deemed catastrophic. Engines, larger boats, new nylon nets, new processing and refrigeration plants for export, the introduction of new plant and animal species, chemical effluents, and other forces combined to deplete fish populations. International aid agencies, local and international compa-nies, government regulators, and local people of many stripes were and are all involved. Both borrowers and lenders have played their part, in large circles and small.

Most dramatically, a giant new fish, the Nile perch (*Lates niloticus*, in Luo *mbuta*), took over. It was first brought from that river to the lake in about 1954, under the British administration's watch, whether for food or angling. This fish, which can grow to about two meters long, multiplied un-controllably by the 1980s, devouring and practically eliminating from the lake hundreds of other smaller species. They ate up most of the local tilapia (*Oreochromis esculentus*), the ngege so dear to Luo hearts, and its at least three hundred related cichlid species in the lake. The Nile perch also gobbled the catfish, lungfish, and other fish commonly eaten ashore—many species of which elders recall wistfully as a bygone thing of their youth (Cohen et al. 1996, Kudhongania et al. 1996). Nile perch itself, oily fleshed, was in the 1980s and '90s eaten only rather grumblingly in Luo homes.

Not so everywhere, though. Overseas, Nile perch was being "discov-ered" and commanded the high prices of an exotic delicacy. Around Lake Victoria/Nyanza ports in Kenya as in Tanzania and Uganda, processing fac-tories, largely foreign financed, sprouted up like mushrooms in the 1990s to cash in on those distant markets. Commercial fishers adjusted their nets and lines for the heavier fish and intensified their deeper-water fishing farther away from shore. Not without protest, though. Kenyan fishery specialists (for instance, Abila 2003) and concerned foreign activists (as in Hubert Sau-per's *Darwin's Nightmare*, a 2004 documentary film on Mwanza, Tanzania and the lake, with Russian planes) publicly decried large-scale international fish exporting that either exploited or just bypassed hungry lakeside people.

Meanwhile invasive plants were transforming the shoreline. The most conspicuous was the floating water hyacinth (*Eichhornia crassipes*), a native plant of South America present in Kenya since at least 1957 as an ornamental plant but observed proliferating on the great lake only since the 1980s. Its growth, at least as remarkable as that of the lantana or striga plants onshore, or that of the kudzu in parts of the southern United States ("weeds," to most Anglophones), immediately affected many human pursuits on the lake. Already by the mid-1990s the hyacinth was choking off large sections of shoreline to boats or nets, and oxygen depletion was creating areas of underwater desert (Kudhongania et al. 1996, Twongo 1996). At the same time runoff of agricultural chemicals like fertilizers and insecticides, and of sewage from towns (all only minimally guarded against either by earthwork conservation or by enforced regulation), was washing nutrients from onshore and stimulating explosive algal blooms. The big lake's phytoplankton mass multiplied at least severalfold in the last three decades of the twentieth century, changing its species composition in the process (Branstrator et al., 1996), and the situation continues. The runoff effect seems to have come not just from the more densely settled parts of the lake's catchment area; it has come at least as notably from the ones less so, where fewer efforts had been invested in terracing and the like. Even though the chemical inflows are less heavy in volume than in some more industrialized countries, this lake's high evaporation and low outflow keep its water and chemical elements contained "exceedingly long, on the order of tens to hundreds of years," making it particularly vulnerable to eutrophication and direct chemical killoff (Cohen et al. 1996, 576 and passim). The results around the turn of the present century have alarmed scientists of varied disciplines, from the lakeside countries and around the world (widely represented in the Johnson and Odada 1996 compendium). Tested and untested, the alimentary, nutritional, and economic effects have been felt wherever people exchanged and ate fish, including all Luo country.

The momentous ecological changes, be they temporary or permanent, have forced more than a few Luo, Samia Luhya, and other lake fishers out of their trade. Others have responded the *opposite* way, by fishing longer and harder. Surely the lean times in fishing have forced *both* sorts to lean harder on their kith and kin and their farming and wage work incomes. A small but undocumented proportion of Luo speakers have coped by migrating to Lake Turkana and the Indian Ocean coast, joining existing fishing enclaves or forming new ones, still partly reliant on contacts back in their western Kenyan homeland for marriage partners, foster care, or sustenance.

Analogously, some of the hundreds of threatened Victoria/Nyanza fish species themselves have survived in swamps and ponds cut off from or only seasonally connected to the lake. Or in the rivers in small numbers. Ichthyologist Leslie Kaufman, a specialist on the lake basin species working with teams of other scientists in all three countries, wrote me in 2006: "River fishing for the once great anadromous runs of barbs (*Barbus altianalis*) and ningu (*Labeo victorianus*) — a principal objective being their tasty roe — are largely a thing of the past . . . recent evidence suggests that these fishes are now largely bound to their river refugia. What we now see for a rainy season spawning run is only a faint whisper of what the grandparents and great-grandparents of today's young fishers knew." What might seem familiar to them is that little native minnow Luo call omena (again dagaa in Tanzania, and mukene in Uganda), still relatively abundant in the lake, and even thriving with few medium-sized fish to be chased by. Humans have filled in that role, though. Luo catch, cook, and eat omena whole, sometimes serving it by the plateful. At night occurs the spectacle of the "city on the lake": thousands of plank boats lightfishing for omena with bright lights and dip nets. For the kerosene lamps, long imported from China, German aid financing has lately substituted solar lamps on credit.

By day too the heat has continued, and for larger fish too. Lake fishers have been using gill nets or long lines baited with endangered cichlid fishes locally called *fulu* or with small catfish, Kaufman observes. People fishing for home included Sunday-afternoon, after-church boaters seeking both food and good company. Mostly they were out for the introduced Nile tilapia (*Oreochromis niloticus* — one of the larger cichlids, which like the still much larger-growing Nile perch is a twentieth-century introduction), or any of a dozen or so smaller native species lately scarce. Elsewhere around the lake, women with basket and cone traps could still be seen wading on the shore, or thigh-deep in the muddy water of the papyrus-fringed wetlands edging the lake, seeking native catfishes and tilapia. But these women were fewer and caught less than they once had (Kaufman, pers. comm.).

A turnaround may have begun, specialists including Kaufman suggest. The Nile perch, that giant introduced predator, was reported waning again under heavy fishing and its own population pressure (Josupeit 2006). Evidently as a result, some of the near-lost species have been seen again in the lake. Leading the way among the fish were native pelagic (open water) kinds, including some of the smaller, herring-like cichlids, some plain and some others brilliantly colored, which like their cichlid cousin tilapia speciate quickly. It has happened as if a Darwinian "nightmare" were flipping to daydream (and regardless of whether ultimately by accident *or* grand design). The future holds no guarantees, but the lake's fish biomass, after one of the most dramatic episodes of species extinctions and near-extinctions ever documented anywhere in the world, appears to have begun—just begun—to rediversify.

Meanwhile, fish have taken to land. Some small-scale farmers around the lake and inland have been trying their hand at aquaculture, especially pisciculture, fish farming. This has come with the technical intervention of American, Japanese, and other international aid agencies (including some of the same ones whose technology undeniably helped cause overfishing in the lake), and with the encouragement of governmental and other local authorities. Faced not only with the lake fish scarcity and broader economic hardship, the rural dwellers concerned, mostly men at first, have literally been digging their way out. They have been constructing postage-stamp ponds with any local hands they can muster, to manage like gardens. This has come about with loans in cash, and recently more in kind, from aid agencies, local cooperatives, and other sources.

The fish farming movement, a local comeback for the cherished tilapia, has gained momentum in western Kenya as elsewhere in the tropics. The unfolding story is testing not just the extremes of fish survival but also the limits of human adaptability and human concern. Although many rural Luo and neighboring people have expressed high hopes since I first witnessed the tentative experimenting some were doing in the early 1980s, they face technical, financial, and logistical challenges, not least so far in the long-distance loans and repayments. Falling as it does in the "liminal space" between farming and fishing (and at the edge of limnology too, indeed), the movement has wide implications that have just begun to be studied. They are not just ecological and economic but also social, political—and not least, since fish are sentient beings, ethical. The ponds may be small, but the waters are uncharted.

Exchange, Conversion, and Cash

On monetization, a topic treated elsewhere (Fearn 1961, Kitching 1980, Shipton 1989), only little need be said here. Cash is itself a sign of trust, however impersonal this might be, since one who receives and saves it must at least implicitly trust someone else to back or honor it. Not always has the nation-state claimed a monopoly on currency or the trust behind it, nor have the nations doing the governing always been the ones who issued the money used. Before European arrival, Luo used various goods for trade—coiled iron wire armbands and hoe blades, or animal hides, for instance—inland and around the lake. Cowries, Arab and Portuguese money, and Indian rupees might have reached inland from the coast before John Hanning Speke in 1858 and later European travelers arrived at the lake. After the establishment of colonial rule at the century's end, a succession of different currencies from Europe, Asia, and Africa flowed into Luo country, and British colonial taxation, only initially payable in kind, pushed Luo into using them from the first few years of the next century.

Money—at present the Kenya shilling—has reached all corners of Luo country but has not taken over the entire economy and I think never will. Nor has its progress been steady or one-directional. It advances and recedes, appears and disappears, as shortages and civil disruptions make barter more practical or appealing, and even in the best times food gifts and loans crisscross the countryside on human heads, donkeys, or bush taxis where roads reach. Cash today is sorely sought and scarce, but quickly gotten rid of, and foreigners are sometimes astonished to see how little of their wealth Kenyans keep in any liquid monetary form, but there are good reasons for this. Cash is a volatile form of wealth, concealable and divisible, and easier than most things to beg or steal. It turns worthless in an era of inflation. It also bears the picture of the president and other symbols of a state authority that has not always been revered in the western hinterlands.

But cash is the medium of the marketplace, and markets large and small have arisen throughout rural Kenya over the past century. Marketplaces are not just about food procurement and trading; they are multifunctional locales. They are the place to exchange news, procure and show off new clothes, and eye potential mates. They are a place to hear political or religious speeches or new music, to gossip, or to joke and banter. Markets are a place to see and be seen. Markets are where to watch the spectacle of whatever funny-haired foreigners are around trading or clicking photos of the brilliant colored wraps, head loads, and little vegetable pyramids that locals take for granted. Bargaining for food in open air, desperate or not, is a sporting game that normally ends with a ritual gift of a free handful from the seller to buyer, to secure a regular customer or just to end the encounter with a pleasant aftertaste. Most established shopkeepers around the edges of markets insist on cash sales, even with known customers. True, there is always a slim chance of a purchase on credit or a cash loan if one can manage to convince the merchant that one's needs are only temporary. But credit from retailers is tight.

Luo thought on barter and market exchange involves some subtle and paradoxical concepts that foreigners may not readily grasp. One of them is a linguistic and conceptual reversibility. The transitive verb *ng'iewo*, for instance, is double-edged: it means either to buy or to sell, according to context. (We see other words like it below, and they appear in some Bantu languages as well.) Another idea embedded in linguistic categories has to do with order and disorder, and I think Luo language suggests a kind of ambivalence about exchanges from one form of wealth to another. *Loko* (v.t.) can mean to sell; it also means to regulate, or, in other contexts, to upset, as one upsets an order.[29] *Rundo* (v.t.), which can mean either to sell food or to buy it, also means to rotate or spin something (cf. Eng. "turn over"). It also means "to mislead or confuse (someone)," or, in the intransitive, "to be stupid or foolish."[30] Linking these meanings may be the idea of "mixing up," and in some contexts there seems to be a negative judgment implicit.[31]

But in other contexts, the idea of conversion is not disparaged. Some use *rundo* to refer to serial conversions of that favored Luo kind mentioned, also familiar from folk "traditions" in other parts of Africa: crops to chickens, to goats and sheep, to cattle, to women (Butterman 1979, 61). Human reproduction completed the cycle, producing more hands to hoe and weed. For men or women, "upward" conversions of wealth to less easily divisible, more animate, and less contestable forms were a saving strategy, among other things, as they helped shelter property from the constant demands of needier neighbors and relatives.

After accumulating some wealth and family by rundo, many farmers have always had to reverse the pattern in distress sales in food shortages and famines, trying to sell livestock (at falling prices) to buy grain (at rising prices), while better-off farmers or traders profited from their hardship.[32] In recent decades, increasing land shortage and rising costs of schooling have gradually changed the pattern of "upward" conversions, making more offspring no longer so clearly advantageous. But something of the value hierarchy still remains, and Luo continue to think it a sign of good fortune to be able to convert grain to livestock, and of hardship to do the reverse. The entry of cash into the rural economy has complicated the value hierarchy but certainly not destroyed it.

Not all that Luo or their neighbors sell should be deemed surplus. Interestingly, much of the maize and other staple foods they sell after harvest they will need to buy back in equivalent later, before the next, and usually at higher prices. Nearly all but the richest do it; and some of the latter profit handsomely by doing the speculating and hoarding for the interim. The seemingly uneconomic practice of "overselling," for most who do it, should not be hastily construed as irrational, as many economic analysts and ob-. servers have done. It allows one selling the chance to turn around and buy something expensive like a farm tool, bike, or tin roof, or maybe even a taxi van at a time when the neighbors and relatives are likely to have food in *their* granaries or money in *their* pockets, and thus to do so without alienating them — perhaps being accused of witchcraft — as one might risk doing at any other time. The seller entrusts his or her food to a system or a process of provisioning common to all, with enough confidence not only that the market or some other source will have food available when needed back but also that there will be kith and kin on good terms in case one needs their help to get it. Even though it can make sense in this social way or others, the pattern may yet be both a cause and effect of poverty, widening a rift between economic classes.

Occupational Shifting and Straddling

The old Luo triumvirate of farming, herding, and fishing have not for many decades been the only Luo ways of making a living. Herding has been constricted by land shortage, as noted earlier, and also in places by land titling and enclosure. At home around Nyanza, farming remains the old standby, but for persons without ample land and lucky rain it stands on shaky legs. Luo and others in their region have long had to look away from the lakeshore, the countryside, and Kenya just to get by.

The variety of jobs they have found or casual work they have invented — both at home and away — is so wide as to defy any description, yet patterns are discernible as the young have followed the beaten paths of their elders.[33] To start with some deemed traditional, healing and divining are possibly the oldest specialized professions in Luoland, as over much of tropical Africa and the world. Home-based beer brewing and (illegal) liquor distilling and vending keep many women as solvent as they are. Blacksmiths, again working at home, form an endogamous caste, or something much like one, here as over much of the continent.

From early colonial times in Kenya, as around British East Africa, the first church mission–schooled young men, followed later by women too, found work teaching, preaching, clerking, and middle managing. Public education later added its own graduates to the salaried workforce. Small-scale shopkeeping and trading, for the early decades of the twentieth century the protected province of Asian-born or descended merchants under the colonial umbrella, have been largely "Africanized" or indigenized as they have spread into the more remote locations. Tailoring, cobbling tire-tread sandals, and tinsmithing goods like trunks and kerosene lamps occupy many hands, especially in the small towns. Animal trading, especially across the Tanzanian and Ugandan borders, has provided great wealth for a few; in a couple of cases I have known it provided them with more wives than they could count on one hand and more children than they could keep track of. Gold digging, panning, and smuggling have enriched the lucky and tempted many others, and work in fleeting foreign-owned mines like Macalder's (near the Tanzanian border, active in the early mid-twentieth century) has likewise employed scores, some brave or desperate souls continuing to dig on their own decades after diminishing returns forced shutdown. In the city of Kisumu and smaller towns that have burgeoned over the past century, Luo have explored a hundred walks of life.

Meanwhile, ever since the British declared the protectorate in 1895, the people now called Luo and the region now Nyanza Province have been known as a "labor reserve." As soon as the first enduring colonial outposts were established in the first years of the twentieth century, locally appointed chiefs aided colonial authorities in conscripting workers for road and railroad building, porter service in the dreaded Carrier Corps, and soldiering—important in two world wars. Forced or voluntary low-paid labor on midland Asian- and European-owned sugar plantations, highland tea estates, and other European-owned farms drew thousands of Luo men, and less often but increasingly women, away from home for seasonal or longer sojourns for the century's first decades. The tea plantations in Rift Valley Province and one sugar factory plantation at Awendo in southern Nyanza remain major draws for work now more optative. Building, factory, and transport work mix Luo and other workers. City jobs as cook-housekeepers or yard workers for expatriates and the rich, and work in all manner of businesses (though only seldom as their captains), have kept Luo in towns even when, or after, they planned to retire back in their old rural homes, as nearly all until lately have hoped to do.[34]

No two sets of figures are alike, but available ones suggest that since Independence in 1963, roughly a third of Luo, Luhya and Gusii men classed as working age, and even higher proportions where settlement is densest, have been away from their home districts at any given time.[35] Labor migrants (perhaps three-fourths of whom are male) appear generally to remit about 10 to 20 percent of their cash earnings; it has been estimated that their contributions amount to between 9 and 31 percent of the average total incomes of the rural recipient households or homesteads.[36] The lives of Luo wage-labor migrants have been intensively studied in Nairobi, Kampala, and elsewhere (Parkin 1969, 1978, Grillo 1973), including back at home once they return (Francis 2000). Many migrants, when away, plan to spend their elderhood home on their farms; but the burgeoning of Nairobi and other cities in Kenya in recent decades suggests that many are starting for the first time actually to stay there for good.[37]

Off-farm work has been a powerful separator of socioeconomic classes for about a century, not just in town but also in the countryside. There a trade, government job, or steady remittance income, perhaps with an added cushion of livestock, can allow investing greater proportions of one's land in risky but lucrative cash crops, particularly coffee (and tea outside Luo country) in the highlands, but also sugar and tobacco in middle altitudes. And these sources together allow richer rural people (along with some urban speculators) to buy up the landholdings of poorer neighbors, including kin, opportunistically in harder times: a process made easier by the national land titling program. Whether class division was a big issue in precolonial Kenya is debatable. Whether it has become one since is not.[38]

All of this has health implications, of course. Among other things, women who are land-poor or landless, or left without expected remittances, must turn to casual liaisons or commercial sex for sustenance more commonly than most would wish. A prime recourse for them is men with big, lumpy cash crop payments, along with truckers and itinerant traders passing through the towns. The risks of HIV/AIDS and other sexually transmitted diseases go without saying in this region, one of Africa's hardest hit.

For most Luo and western Kenyan people, work and family economic life are about shifting, straddling, or combining. Economic decline from the 1970s into the new century has made this even truer than before as wage jobs have become scarcer and more temporary. Just as keeping scattered fields helps spread out the risks of farming, keeping more than one occupation helps soften the blow when any one of them falls through. Basket weaving, quail keeping, and countless other hobby-like cottage industries help fill in the cracks of livelihood. "Occupations," often as not, are plural ones; and remittances from kin are vitally depended upon if not often reliable, particularly for women left at home to manage farms and families on their own. Migrant husbands' failure to find or hold down work, or to remit from what they do earn, gives women both added burdens and probably also—because of their greater contributions—extra voice in their home and neighborhood affairs. This is so even where men keep claiming control over the biggest farm decisions from near and far, as remains the norm.[39] The striking irregularities of income make borrowing and lending a crucial strategy of survival in Luo country, despite sobering class differences. Knowing that it will be needed helps keep the social tenor and etiquette of the countryside as agreeable as it is.

Interethnic Relations: Luo and Their Neighbors

Kenya is not an ethnic melting pot; a centrifuge would be as apt a metaphor for the past century's spinouts and separations. Its people have tended to live in rather clearer ethno-linguistic blocks than people in some other parts of Africa.[40] The Luo population, in its bulk, is no exception. As speakers of a Nilotic tongue, Luo are isolates, or nearly so, in their present Kenyan home area, migrations having cut this off from those of their closest linguistic kin, the Padhola, Lang'i, Acholi, and other groups in eastern and northern Uganda. To the north of the Kenya Luo heartland, about two million of Kenya's roughly 3.1 million Luhya-speaking people, a Bantu-speaking amalgam of diverse subgroups, predominate in the hills of Western Province.[41] To the northeast reside Nandi and other members of the Kalenjin group (linguistically often classed as "eastern Nilotic" in origin; Luo are "western Nilotic"), another ethnic amalgam only a few decades old. To the southeast live two major Bantu-speaking groups: the territorially compact Kisii, numbering about 1.3 million in 1989, and further south, about 97,000 Kenyan Kuria who, like Luo and Maasai, have significant counterpart populations over the straight-edge border in Tanzania. A thin piece of Kenya's extensive Maasai country abuts southeastern Luoland between the Gusii and Kuria home areas. Over 108,000 Luo-speaking people called Suba or Luo-Suba, former Bantu-speakers or their descendants who probably share close ancestry with Gusii, live divided into two enclaves on the lake and its islands, and in the southeastern corner of the Luo country near its Kisii, Maasai, and Kuria ethno-linguistic frontiers. South of the Luo population in Tanzania's northern Mara region live a diverse mixture of Jita and other Bantu-speaking peoples.

In what Luo call their own country live immigrants and others who do not call themselves Luo, and cultures have blended in some ways but not others. Aside from assimilated Suba, splinter groups of Luhya from origins in Maragoli and other crowded areas to the north constituted, at just under 100,000, the biggest ethno-linguistic minority surrounded by Luo in the 1989 census; most of these or their forebears had moved to southern Nyanza for more land since tsetse clearances in the 1930s and '40s. In most material ways, at least, these people live much like Luo. Intermarriage between Luo and their Bantu-speaking neighbors is not unknown, but it remains the exception.

In Nyanza Province's various resettlement scheme areas, including its formerly Asian sugar plantations, and elsewhere, some 8,700 people classed as Kikuyu, 6,500 "Kalenjins" (including Nandi, Kipsigis, and others), and several thousand others from central or eastern Kenya lived among Nyanza's Luo, Gusii, and Luhya at the time of the 1989 census.

In Kisumu — Nyanza Province's capital and Kenya's third largest city — and in other smaller towns, a few thousand people of Indian, Arab, and other trading minorities from afar have flourished since early European times and before.[42] Most distinctively visible among these are Indians, who hail largely from Goa and Gujerat states; Somali, who tend to concentrate in restaurant business; and British and other Europeans, who divide between commerce, aid and relief work, and government advising and consulting. Few of these people, particularly among Indians or Arabs, have intermarried with Luo or other populations deemed indigenous.

Meanwhile, thousands of Luo people themselves have traveled widely and have colonized sizeable neighborhoods of Nairobi, Mombasa, Kampala, and several East African cities and a few isolated rural areas outside their lake basin heartland. Luo rural-urban migrants tend to move in first with kin and friends who have preceded them and help them get started.

Relations between Luo and neighboring groups in Nyanza defy easy generalization, having changed character over both the longer and shorter terms along with shifting currents in Kenyan and regional politics.[43] Luo, Luhya, and Gusii have tended to remain on quite amicable terms through most of the past century, though not without stereotyping and misgivings, or the occasional boundary skirmish. This relative harmony is easy for foreign analysts to overlook — "tribal clashes" always make better headlines — but something not to be taken for granted, given the sharp Nilotic/Bantu linguistic divide and given the extremely high densities of rural settlement these groups have reached. There is a history of *intra*national African diplomacy to be written on western Kenya.

Luo and Nandi have a rockier history; stories of older generations' cattle raids have found new, conscious parallels in armed, explicitly ethnic clashes in the early 1990s over intersettled lands at the interstices of these groups' territories. Kuria people trade with Luo in towns but otherwise have very little friendly contact with them, the two groups deeming each other old cattle-raiding enemies. Maasai and Luo ditto. Oral histories tell of a silent trade—exchanges based on trust without meeting—between Luo and Maasai in the late nineteenth or early twentieth century, but probably only in recent decades has the commerce in traded goods between Luo and Maasai exceeded the commerce between them in raided or stolen cattle.

Gikuyu, Meru, and Embu, together that great central Kenyan population that has dominated so much of Kenyan metropolitan and political life since early colonial times, remain too far away across the great Rift Valley for Luo and them to raid each other's cows, but their historic competition with Luo has extended to other domains in towns and other public settings. Economically Gikuyu and allies have usually come out well ahead of the nation's geographic marginals. Despite resentments, their leaders and those of Luo, Luhya, and Kalenjin have at times formed political alliances of expediency, their configuration shifting with every weekly news magazine and always fascinating if threatening the nation by its instability. None of these rivalries qualify exactly as "ancient tribal enmities" (they soften too often for that), and politicians hitherto have constantly denounced "tribalism" for national unity here as elsewhere in Africa, but Kenya remains a nation with deep ethnic divisions—variously conceived as racial, linguistic, cultural—no less.

Luo Identity and Fit

A few remarks remain to be made about the Luo people and their feeling about their place in the world. It is well to ask here straight out: Who, or what, *is* Luo?

A Luo is someone who considers him- or herself a Luo and who is so deemed by other people; but even to say this is to place more emphasis on individuality than may be due, since to be a Luo it can be important to be (or to act) among other people so considered, and thus a part of something bigger than oneself. There are many specific traits sometimes used to peg Luoness. It can be a name beginning with O (for a male) or A (for a female). It can be a look (to some, a "race"), a language (one with flowing, musical sound), a store of songs and proverbs and dances, or a proper respect for seniority and precedence. It can be sporting a spear or shield of a particular design, or a rarely displayed ceremonial costume with hippo tusks and ostrich feathers to "savage it up" (something startling or bizarre even to "Luo" themselves)— with or without sunglasses. Up to a recent era it could be a special process and pattern of tooth extraction from the lower front jaw. Still now, it can be a growing up without genital cutting. Or, again in our own times, it could be a crisply pressed shirt donned on a visit to town, a vote for a Luo candidate, or a doctorate of philosophy. But there are snags. More than a few Luo define themselves as Luo because they are fish eaters but just now have no fish to eat; and others know they are not Luo if they circumcise but are being rightly or wrongly advised in one clinic or another that they should.

Of all the indicators just mentioned—and the list could go on—it is best for most purposes not to pick out any single one as a sole determining criterion, because that is likely to lead only to stereotyping and division. Luo borrow and lend identities, like anyone; and the package of Luoness often comes incomplete. To a degree their physiognomy carries messages, genetic or human made, but they do not always wish to be locked in.

Representation of identity is both a personal and political issue. Not everyone who is or is called Luo, in whatever of these ways, would wish to be known by this as a first or main identifying feature. Some prefer to be Kenyans, Luo Tanzanians, Ugandan-American Luo-Luhya, East African Muslims or Christians, Luo people of the world . . . or just people of the world who have long since buried everything having to do with "tribes" and tribalism. No one, in the lake basin or anywhere else, ought to have exclusive power to define another for good, or be compelled to accept one definition or set of prioritized identifiers for all contexts. Particularly not since the Luo country has in it many from outside—traders, refugees, missionaries, anthropologists, and aid workers, for instance—who have gone part native, other erstwhile insiders who have partly decamped from Luo roots, and still others who have intermarried or been born of mixed parentage. Ultimately we humans are all of mixed origins.[44]

Being situated at the geographical margins, and at times the political ones, of the Kenyan nation since well before its independence in 1963 has left Luo with lingering questions about whether they belong in it at all. Being Nilotic speakers outnumbered by Bantu-family language speakers in Kenya adds to that feeling while encouraging recourse to English. Forced conscription and not fully requited taxation in colonial times, and job discrimination in independent ones, have scarcely been forgotten; nor have occasional incidents of mysterious assassinations of important Luo leaders (most notably statesmen Tom Mboya in 1969 and Robert Ouko in 1990). Luo people attend the government meetings (*baraza*), with their pep rallies and fund raisers, and chant the chants like everyone else, if only out of fear of the consequences of not doing so (see Haugerud 1995 on Embu).

But few are fooled. Sporadic incidents of crowd anti-government violence, especially in Kisumu, about various causes (sometimes perpetrated by government army and police guns) have given the lie to all that in the late twentieth and early twenty-first centuries. On a daily basis, the resistance and subversion have been quieter and subtler: evading official buyers and farm regulators, letting loans go unrepaid, and so on. Luo as a whole have never really entrusted their lives to any national government, nor indeed to any other centralized human authority.

Luo people have tended to feel more at home in their Christian and hybridized indigenous churches since the first missions were founded in the first decade of the twentieth century.[45] Early missions carved up the western Kenyan countryside into territories for hegemony, and by and large they have respected each other's turf. But indigenous faith and practice have woven in richly with each, and some have hived off independently, so the end result has been a plurality of variously or at least nominally Christian religions in every locality. The local churches—built as they are on neighborhood, including matrilateral kin ties—grow and multiply much like lineages but often afford special status, role, and voice to women as patriliny and virilocality together hardly do. Universalist ethics, where espoused in church, need not clash with these local roots but instead often seem to complement them. If churches were accompanists or instruments of colonial rule and agents of incongruous cultural impositions and wrenching reforms in living style, they have as often in Nyanza, since earliest colonial times, been seen otherwise as a bulwark against state authority, or a place to escape, rally against, heal from, or transcend its more brutal effects, and the effects of poverty and illness. The borrowing, lending, tithing, and gifting that go on in and around churches and their aid agencies—licensed or not by other branches of national government—make up a vital and adaptable part of this action.

Luo people have often been portrayed as an egalitarian people (see Odinga 1967, Parkin 1978). In some senses this is quite true. The ideal of equality (*romruok*, lit. amplitude or parity, or *winjruok*, lit. understanding) is evident for instance in the diffused nature of traditional political authority, the help lent or given to hungry kin and neighbors after a cattle disease or a crop failure, the lodging afforded home-area migrants in towns, and the food-sharing ethos that extends all the way down to (and maybe rising up from) very young children. Individuals who get too rich too quickly are subject to neighbors' accusations of witchcraft, while those who suffer sudden misfortune are suspected to be victims of it. Everyone is expected to have a chance in life to subsist, to marry, and to reproduce: the most basic human rights and responsibilities in East Africa.

In other ways, though, Luo are decidedly and extremely *in*egalitarian. The generations-long memory of who is a belonger on the land (wuon lowo) and who is a "dweller" or land client (jadak) pervades rural social relations. Subtler is the habit of referring to married women by their birthplaces ("Nyalego," daughter of Alego) as reminders of their insider-outsiderhood in their conjugal homes, even though this is not usually meant to disparage them. Richer people, and ones with bigger families, have far more say in Luo local affairs than poorer or less prolific ones. The hierarchies of generation, birth order, and marriage order are all but absolute in Luo ideals, and constantly played out in the etiquette of deference and avoidance of alternate generations. The worst abomination in Luo country is to forget one's place in this chain of being.[46] Bitter dramas ensue when, say, a son, having grown richer than his father, denies his requests for help or beats him; or when a young married woman talks back to, or raises a cooking spoon at, her husband's mother. Usually it does not come to this, but keeping order takes work. Luo and western Kenyan life are full of asymmetric expectations and demands for animals, work, food, and money—as gifts, loans, or simple requisitions.

Neither equality nor inequality, then, sums up by itself the Luo way of relating to others. Luo people know they are equal, if not superior, to any human alive; but they also know they occupy a place in a moving order. If they play the game right and respectfully, they expect, as elders and progenitors of large families—and then as ancestors of long lines to remember them—eventually to get their chance.

Learning in Luoland

Now, before getting deeper into the ethnography, the customary brief word about methods, to pick up the story left off in the introduction and make a suggestion or two in case they might be of use to others in the future. One does not want to be a stranger in western Kenya; almost any contacts are better than none. Another sort of professional insider-outsider, a missionary of decades' standing to whom I had been introduced by correspondence, helped me make my first contacts on arrival in the Luo country. Asking him to do so was a gamble on my part if not also on his. A missionary could be cherished as a healer, teacher, and spiritual guide; resented as an arrogant cultural imperialist who would never understand initiation, polygamy, or home burial; or just mistrusted as someone who raised hopes with big promises. Fortunately this one turned out locally beloved.

Whatever they say about individuals, Luo people pride themselves on never turning away a traveler in need of hospitality. Indeed, in one legend, villagers who refuse to let a wandering destitute old woman stay and enjoy their bounty and celebration are cursed by her magic and flooded out of their home (the origin of the lake Simbi Nyaima—transcribed in Onyango-Ogutu and Roscoe 1981, 138–39). Whether because of such warnings or not, Luo are on the whole an accommodating people (doubtless a key to their expanded numbers in the past). If anything, the visitor in Luo country may worry about *too much* hospitality: about having animals slaughtered one would rather see spared, or being given a live hen when departing from an overnight stay. All part of relationships and reciprocity, a game western Kenyans understand better than many of their visitors.

For me the most important "research method" of all—more important than surveys or anything—was from the start to make friends and trusted acquaintances who would mean more than just research. Of course, the ones first willing to latch onto a newcomer were not, for the most part, the ones who later emerged as the truest friends, nor were their ethnographic statements always the truest stories. A few of those people turned out to be ones whose craftiness or suspicious histories had marginalized them from the community, which was what had made them so available in the first place. But others were just temporary residents, like schoolteachers from other communities or ethnic groups, who were eager to get to know another in a comparable position. Finding the local people worth relying upon took time, intermediary contacts, and serendipity.

A second important "method" took more practice and sometimes forgetting the clock: to combine active learning with passive—including just walking or relaxing in good company and absorbing their concerns. Participant observation, that favored hallmark method of anthropologists, describes much of what I and we did around and about (though sometimes just as participant and other times just as observer): trying my hand at plowing or hoeing with a short-handled hoe, strategizing with neighbors about a marriage, attending funerals, and the like. Friends merged with teachers of local ways, though there were some who were just one or the other, and it was usually not hard to keep things quite congenial, if not always high spirited in mood — for whatever counts as "field work" always has its ups and downs, and going through those increasingly in synchrony and empathy with those around, as "field" becomes more like home, is part of what gives it its meaning.

The elders found me convenient, inexpensive accommodation in a vacant country shop, or elsewhere a school compound, or better still, kindly lent a little house in their own homestead to sojourn in among the daily activities of their family. The hosts and friends, including neighbors at times or a hired local helper-turned-confidant at others, always made sure I got fed with Luo or other Kenyan food (wholesome and to me usually tasty; though not spicy by world standards, it is carefully and agreeably textured); and on occasion pricier foreign delicacies they expected I would enjoy as a touching gesture.

After a year in western Kenya, as mentioned, I was joined in Kagogo for extended intervals by my spouse, who had come to Africa to join me and to work as a teacher and editor in Nairobi. She too formed strong attachments in the rural setting where we lived, and she accompanied co-workers and me on some of our travels around. Her participation in many aspects of local life, as well as willing or inadvertent participation in my own work, enriched both the work and the experience much.

For help in interpreting or in repetitive tasks, I chose mainly young adults from the communities who had been recommended to me by their former school headmasters or by elder neighbors. Several young men and a young woman, having left school or in two cases college, worked long hours with and without me, mostly in ones or twos. Far more than just assistants or co-workers, some became fast friends, soulmates, and key informants too, and the ties to kith and kin they opened in their communities proved crucial.

Studies for which I worked with these local co-workers and helpers (whom I trained, while in other ways they were training me) included a basic census and genealogical survey of 286 homesteads in Kagogo, for about 95 percent coverage. We followed up with a deeper survey on selected topics (related to land, agriculture, livestock, and exchange) with a subsample of 107 of these homes, again in a contiguous area, covering a valley. With other assistants and a different, stratified sampling method, I surveyed representatives of a total of 164 homes in Kagan, Uyoma, and Isukha, mostly about matters related to their participation in aid programs.[47] These surveys were designed only after months in western Kenya, and in each survey we interviewed males and females, in roughly equal proportion, in each locality. Conversations and interviews—of all levels of formality and informality—were conducted usually in people's own homes, with or without interpreters, alone or with onlookers—or with friends in our own home. The surveys took us much time and effort, in both information gathering and coding, and yielded far more information than I have been able to use. We saw much that we were not looking for, and more than a few people we met this way became valued friends or acquaintances. But survey methods in rural Africa have their drawbacks and pitfalls—it is a delicate task to fit them to local language, etiquette, and values. I caution anyone who wishes to know rural Africa against relying on survey data alone.

As for tools and trappings, I had feared that our having access to four wheels at times might set me too far apart from people who traveled on foot, but it turned out as often the opposite. The seasoned, thornbush-scratched all-terrain vehicle offered to me by a paleontologist overseas on long-term loan, and the smaller, more fuel-efficient one I later bought—also well used already—scarcely moved anywhere without passengers, sometimes interviewed along the way. Being able to drive sick people or pregnant women to the clinic, or haul cement bricks or roofing from town when asked, helped in making friends. The periods without wheels, and the trips crowded into bush taxis, presented their own different learning opportunities. Camera and tape machine got lent around increasingly, for better or worse. I brought them out less than some; until I knew people well such hardware felt either ostentatious or obtrusive. During or after interviewing I relied more usually on pen and small notebook. Then there were the times, often the best or worst, with no recording devices at all.

A word on language. People in Kenya, and Kenyans abroad, have happily taught me their tongues since before I arrived there, not least the tireless children. That English (my native tongue) is spoken by the older ones in schools though, and that schooled Luo are comparatively good at it, adds some challenge for even a determined Anglophone seeking to learn their language well. In country, missionary tapes, then rare, and a grammar book first helped launch me in (Dho)Luo to overcome that.

In the case of the East and central African lingua franca, (Ki)Swahili, also taught in Kenyan schools, I had some training before going to Kenya and some further classwork and books to help me while there. I found, somewhat reassuringly, that as far inland (or "up-country") as the lake basin, others simplified and mangled this coast-rooted language pretty well too. (Few, for instance, used more than a handful of its more than a dozen noun cases, let alone making adjectives agree.) At times it was handy to have a neutral language to turn to in which no one had the upper hand or the temptation to keep correcting: a different kind of understanding.

Within a few months I could get around without much help in both the tongues I was learning, and I continued eagerly to learn in every stay. To imply perfect knowledge, though, as (too) many ethnographers have done, would be to commit injustice to these ancient, complex, and constantly adapting languages, rich not only in meaningful tonal inflections (in the Luo case) but also in proverb and allusion. Hence, even once able to converse my way through most situations, I still used and appreciated help often from assistants. For (Lu)Luhya, spoken in Isukha, where I sojourned several brief weeks, an interpreter was indispensable.[48] Altogether the notes I took in English contained more and more Swahili and then more Luo terms and phrases in it for convenience as the untranslatable became more evident. Living and working in more than one language, one more easily than another, gave me something important in common with Kenyan people themselves. In East Africa the switching and interpreting are part of life's fun.

Loans in Luo

Language provides a way to begin to understand how Luo people think about loans and entrustments of the things around them. The Luo tongue, DhoLuo, suggests that credit and debt are not necessarily conceived of as economic conditions, or as opposites. *Holro* (n.) is a loan, and *holo* (v.t.), to lend (a term that takes the person lent to as its object), also means to borrow (something).[49] There is something like a reversible perspective in the term, as there is in *ng'iewo* (v.t.), which can mean to buy or to sell.[50] The concept of *holo* might better be understood as "to engage in a loan (with someone)," as *ng'iewo* might be understood as "to engage in a sale." *Holo* can take both a direct and indirect object, but where there is only one, it can be the person(s) rather than the thing. This parlance underlines the social dimension: lending is not merely something one simply does to a thing, but a way of relating to one or more people. *Holo* may be a cognate of *olo* (v.t.), to flatter. What lending and flattering have in common is ingratiation—both are ways of obligating or gaining power over someone. To ask repayment of a debt is *bandho* (v.t.), best translated as "to dun"; again the usual object is the person(s).[51] The language of transactions puts the emphasis on the social dimension.

Many things, economic or otherwise, can be expressed as debts. To speak of repaying a debt (*gowi*, n., pl. *gope*), Luo use *chulo* (v.t.), which can also mean to pay (as in a sale or a fee payment), obstruct, prevent, clog (as in filling a hole), or to suffocate (cf. Eng. put paid to, or amortize, from a word meaning "to kill").[52] The sense of these multiple translations appears to be that a repayment rights something wrong or missing—as when Anglophones speak of "settling," "squaring," or "clearing" a debt or a breach. One can repay a good or an evil, as in English—*chulo kwor* is to take revenge. A satisfactory ending for a loan is a kind of *romruok*, equality (of payment and repayment, or of persons) or sufficiency.

Debt can be variously conceived of as a state of being, as in the English "to be in debt," or, to distance it from the self, as a kind of adverse possession. *Bedo gi gowi*, to owe, is literally to sit, rest, or remain with debt, and "I owe you" translates as *an gi gopi*, literally "I (am) with your debt" or "I and your debt" (are together). This same construction is used to signify ownership or possession, as in *an gi dhiang*, "I have (or own or possess) a cow." "I 'hold' or 'carry' your debt" would be the nearest figurative translation. An etymological link between owing and owning appears in English, too, as suggested earlier; but it is clearer in contemporary Luo parlance.

The broader concept of "entrustment" that I have suggested encompasses lending as one of its many forms might be best translated by their term *geno* (n., v.i., v.t.), to hope, trust, expect, depend on, or have confidence in; and *genruok* (n.), confidence or expectation. Some entrustments are discussed without even any word like *holo* or *geno*, as when *riembo* (v.t.), to send or drive away, is used for entrusting a domestic animal to someone outside one's home, with an expectation of return.

Though Luo have terms of their own to mean "loan," when speaking about official credit and debt (as when they borrow from government cooperatives) Luo most often import the English term *loan* itself into their DhoLuo utterances. Their use of the foreign term is important. It suggests they perceive these credits and debts as alien or colonial contrivances rather than as something of their own making.

Significantly, too, the Luo language appears to have no word exactly corresponding to loan "interest." (About the nearest would be *ohala*, used for *any* profit or gain.) Elders insist that loan interest was an alien idea before the government and other "foreign" lenders began lending in their country, and I have scarcely heard Luo use the English term for anything but financial dealings, and official ones at that. I once asked a thoughtful farming man in his fifties, under the thatch of his Kanyamkago home, whether the Luo custom of lending a cow and collecting more than one offspring calf from the borrower — a custom familiar throughout the countryside — was anything like money lending at interest. At first he looked puzzled. He thought a moment and then nearly fell off his seat with laughter. "Yes," he finally said, "I suppose you could say it is." But he was still grinning.

The exoticness of the idea of loan interest in the Luo country is one reason why credit schemes of many kinds have run afoul in the rural areas. Nor is Luo country unique in this respect: across Africa, northern and southern, Islamic and Christian, one can find intercultural misunderstandings about whether time is money, and whether it is fair to let interest mount indefinitely, in linear or parabolic progression.[53]

In treating the complex subject of loans and entrustments, an Anglophone is tempted to lean on the language of markets and money, if only to simplify. I shall suggest that some of the most meaning-laden occurrences in human life — among the Luo, transfers of property upon marriage or homicide, or certain kinds of labor sharing, for instance — are in some sense entrustments or loans, since they involve reciprocities over lapsed time. But they are not *merely* loans, and a term like *loan* or *credit* scarcely suggests their richly layered meanings. Thus to reduce flesh, blood, and sentiment to a measurable quantum is unnatural, and it obscures as much as it reveals. It also raises questions of perspective. When Rispa Nyayal of Kanyamkago, for example, is married on condition of later reciprocation in cows and cash to her natal kin, she may feel, quite justifiably, that her marriage is not a "loan" at all. A word like "resource" suggests something teleological: that it exists only to be used, or perhaps used up. This raises moral problems where labor is concerned, and when animals and children come into play as entities that can be entrusted and exchanged.

This does not mean one must expunge all market language from discussion, for to do so is to enter a game of "John Brown's body" in which all our terms disappear one by one. We need the words we have — but we should not import them uncritically. There are some narrowly economic senses in which transfers of living and breathing persons, and animals, have directly (and successfully) competed with loans of fertilizers, seeds, and cash. Things occupying quite distinct cultural and symbolic spheres have been co-mingling within an economic and political sphere. They are both alike and not alike.

1. Boy, his entrusted herd, and a Luo homestead in background, on flatlands near lake. All photographs by author unless noted.

2. A child and her charge, in Kanyamkago. A common sight in western Kenya. Photo by Polly Steele.

3. Lighter in company: girls carrying water home.

4. Young men plowing fallow land, Kanyamkago.

5. Woman digging with short-handled, iron bladed hoe (kwer), Kanyamkago.

6. Two neighbors and friends, with granary in background, Kanyamkago.

7. Boat for fishing and ferrying, Winam Gulf.

8. Marketplace at midday, Migori town (Suna).

9. Women with head loads, Migori market.

10. Patriarchy and (for some) polygyny, concealing subtler female power and influence.

11. All-female, three-generation home in Isukha (in Luhya country).

12. Casual teasing, more common between alternate than adjacent generations. Kanyamkago.

13. Two milkings next to a house and a cattle enclosure, Kanyamkago.

14. Elder woman mincing sugarcane. Some distill to sell to invited drinking parties.

15. The rural reach of the state: chief with elder, Kanyamkago.

16. Stirring maize kuon (Swahili: ugali) at a funeral, Kanyamkago.

17. Dancer with ostrich feathers, cross, and spear at buru ("ashes") ceremony, Kanyamkago.

18. Cattle chosen and garlanded for sacrifice at buru, Kanyamkago.

19. A custom under question: widow and "guardian heir" (kinsman of husband), Kanyamkago.

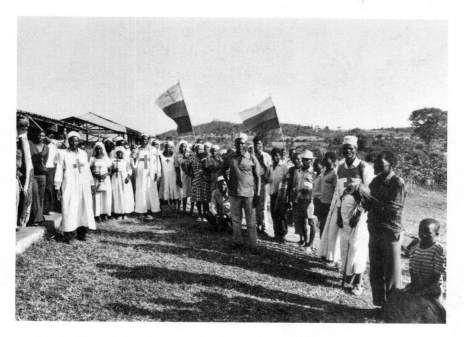

20. A Roho (Spirit) church group assembles for a march.

CHAPTER 4

Entrustment Incarnate

Humans and Animals over Years and Generations

A man without debts is a man without friends.
—WOLOF PROVERB

W ho belongs where, and among whom, is a complex question in East Africa, even for babies and young children. In this chapter we begin to look into the movement and sharing of living, breathing beings: transfers and counter-transfers of humans and animals over time. These are loans in a sense, but seldom just that; here one enters a zone where economic terms like loans or debts become hard to apply, or seem too simplistic, on their own, but where entrustment and obligation are no less vital. Exchanges of sentient beings as described here are not merely contracts entered into by individuals but structures and processes involving groups, networks, and categories of persons with varied and overlapping interests. They are not just active, voluntary agreements but also circumstances into which people can be born, and conditions under which they may even live their whole lives.

Farming Children Around: Child Nursing and Fosterage as Entrustment

Children in East Africa have traditionally done most of their growing up in the company of other children, including ones older and younger, and the customs and arrangements by which parents entrust elder children with the care of toddlers and infants are an important feature of life in the region. It is striking to visitors from overseas to see everywhere young girls of six or eight years old carrying babies slung on their backs in khangas, at home or in the fields while their mothers work. Much child nursing—children caring for children—occurs within a household or homestead group, and some between groups. It could hardly be otherwise in a region like western Kenya: a region where birthrates are high; where men are often living away; where a family's fields are scattered and much farming done at a distance from home; where farming, shopping, and cooking are so labor intensive; and where the division of labor so skewed anyway as to keep women busy all day long. Parents, and mothers in particular, do not have time to care for as many children as they have, over as many years as they may have them. So they delegate elder ones to look after their juniors. They do not, at least at first, just leave them on their own. They coach the child caregivers and provide them backup for emergencies, either in person or, when they are away, by proxy through other adults around.

A girl as young as three or four begins to imitate older children in caring for infants. She may be assigned a boy or girl, but a boy is likely to be given only another boy to look after. Often a first child is given a third, a second a fourth, and so on. *Jopidi* (sing. *japidi*), or child nurses, are at first given some parental oversight but by five or six may begin to take fuller charge. From then until about ten or eleven years old, they carry around younger children on their backs, and rock, wipe, and sing to them. They try to keep the children out of trouble with the animals or their droppings, or the *panga* or machete. In some cases they torment them, but in my observation, tender care seems to be more the rule. The big risks, rather, would seem to be risks of error: of a knot that comes untied and lets the baby fall from the back, or of a moment's inattention that lets the baby crawl onto the hot kitchen hearthstone—neither of which have I ever seen happen, though happen they must. Jopidi take license to discipline the younger children in their charge, whether by a rebuke or an occasional slap. Jopidi do practically everything with, and for, their charges, and in the worst cases to them, until the young ones are as old as five or six and perhaps able to take on still younger children, becoming their jopidi in turn. This is not just babysitting—that term is far too passive—but nearly total caregiving and surrogate parenting. Seen the other way around, parenting is an adult version of what most everyone has known how to do since late childhood.

To look upon children as active providers of care, rather than just as passive "sitters," opens one's eyes to some subtle flows of benefits "upward" as well as "downward" through the generations. If children, by helping care for other children, not only free up their parents and other elders from labor but also help maintain the system of intergenerational "social security" that will maintain the elders in the future (as well as themselves too, in time), they play a more central role in economic life, and in society's continuance, than often credited. The flow of care and milk, even in contingency, is expected to be a structured flow. In all this there is plenty of sharing, but it is not just free, aimless sharing. It is culturally programmed but also actively managed by elders and peers, in practically all the activities of home and farm life. Distinctions blur here between adulthood and childhood, between work and play, and between what is economic or educational and what is not.[1]

Children move around fairly frequently between homes, for visits as short as a meal or snack or sojourns lasting months or years. Just how frequently no one yet has accurately measured, but one on the ground need seldom look beyond a few homes away to find a case. In some respects this treatment of children parallels the treatment of large animals lent between "stock associates," to be described. The custom of fostering children away is not usually spoken of in the same idiom as sending cattle away (*riembo*)—a noun like *pidi* (nursing) or a verb like *pidho* (to plant) is more common for fosterage. It is not polite in East Africa to compare human beings to animals too explicitly without explanation (safer the metaphor of plants). Be that as it may, some of the underlying reasons why it happens are similar. It happens in part because fertility and childbirth rates vary widely if sometimes inexplicably from one woman and family to another (as animal fertility rates also vary) and some homes have many more young children or animals to tend, or fewer hands to tend them, than their kin, friends, or acquaintances do. This imbalance in turn comes about partly because families at different stages of their developmental cycles experience different needs for working hands. With animals, it also comes about because of differing needs for animal traction or milk. With both child and animal entrustment, who asks for the arrangement and who benefits vary from case to case: it can be the "lender" who is getting the main benefit by being free of a mouth to feed, the "borrower" who gets it by acquiring the use of labor power (and with lactating animals, milk), or both. In the cases of children and of animals, lending also makes or breaks social ties. One final comparison to make is that both children and animals are sometimes abused when entrusted in foster care—probably more than at home—and both may well feel it when it is happening.[2]

A difference is that with humans, the entrustment often involves class differences, and the children themselves may be learning to feel these as much as their parents do. It fairly frequently happens that one household with many children but little to feed or clothe them fosters one or more to another, economically better-off or higher-status home in country or town—where their duties are likely to include helping care for children younger than themselves. Children also go to visit grandparents, who on their mothers' sides are likely to live at a distance; and they often develop special sympathy, camaraderie, and humor with them—as when a girl and her grandfather refer to each other as husband and wife—that can contrast sharply with the more formal, deferential relations they are taught to develop with their own parents.

Children's movement between homes, and between parents and siblings for care, can be considered fosterage, but because they often serve as active workers as well as dependents, and learn new skills, one could also call it a kind of apprenticeship for adulthood. Sometimes an older sister who marries out of her home takes a younger sister or cousin to help her in her new home. Eventually, if she starts behaving like a junior wife, she may even become one in a more formal way (Wilson 1968, 123, para. 109).

A system of classificatory kin terms, which equates all cognatic ("blood") kin of the same sex and generation by identical terms of address and of reference, fits the pattern of children's frequent moving between homes. It semantically draws kin together so that, for instance, what would be "cousins" in English become brother (*omera*) or sister (*nyamin*) in DhoLuo, what would be "aunt" becomes another mother (*min*), and what would be "niece" becomes a son (*wuod*) or daughter (*nyako*) of one's own. Moreover, the usual use of plural possessive forms such as *minwa*, meaning *our* mother, or *wuonwa*, *our* father, customary whenever addressing or speaking about members of ascending generations, further unifies the speaker with all those implied to be included. When one moves to the home of a parent's sibling, or even to the home of what would in English be a parent's first, second, or third cousin (and so on), one continues to apply the same terms as those one used in one's own household. It is as if the terminology system and conventions of etiquette—alas, obscure in their origins—were designed to make it easier to foster children around—or just to reflect such a custom already in practice.[3]

Part of the early training that children get from their mothers, and from foster parents and child caregivers alike, is the free and frequent giving and taking of coins, toys, or snack food, and their passing on in turn to younger children. They also learn not to steal from kin, a lesson they will need to apply later in life.[4] In these things there is a moral relativity: theft from kin or clansmen is a moral failure; raiding from strangers, especially from members of Nandi, Maasai, or Kuria, other ethnic groups, is more likely to be deemed a morally neutral act. The neat schemas of concentric circles of Evans-Pritchard's Nuer text or Sahlins's sociology of "primitive exchange" cannot be neatly transposed to the Luo case, for they take little account of fortuitous friendships or of generous or stingy personalities, but skewing of favor is certainly recognizable. Church leaders and schoolteachers have sometimes tried to impart ethics of universal fairness and kindness. I doubt that these ideas were entirely new to Luo when the outsiders arrived—some of the early "explorers" and missionaries were treated too well for that to be true—but they have not been the prevailing ones in the past, and whether they surface at all doubtless depends on the particular situation.

Having been cared for by an elder sibling or cousin is likely to leave both parties with a lasting sense of a debt or obligation, and a clear assumption of hierarchical responsibility. Having sojourned in more than one home growing up entails responsibilities, and expectations of reciprocity that may extend a lifetime, or a serial passing of favors down through the generations, either directly or even, more importantly still, by leaps to alternating generations. The exchange of children in visiting and fosterage contribute to the mechanisms of insurance by which Luo and other East African people safeguard themselves from food shortages. In famines, girls are sometimes pledged away by their fathers for eventual marriage to their hosts and benefactors, as noted elsewhere. The alliances cemented, even where famine and marriage pledges are not at issue, may prove useful later in times of crisis.

An orphan is likely to be fostered into the home of a woman who is childless (n. *lur*, especially if young; adj. *migumba*, especially if elderly or lacking a male child), one too weak or disabled to handle domestic tasks, or just anyone who strongly enough wishes more company. This can be a great solution to two problems or a whole new set: these relations appear variable in quality. At best, the foster child (*misumba*, pl. *misumbni* or *wasumbni [a Swahili-ized term]*) is eventually treated as an ordinary family member—that is, the genealogy simply gets adjusted for all practical purposes. At worst, the arrangement becomes a form of indentured domestic servitude subject to abuses reminiscent of Cinderella and her wicked stepmother. The latter outcome has been common enough that Luo lexicographers and others have sometimes translated fostered *misumbni* as "slaves." I caution the reader that both folkloric analogies and especially translations like "slave" are likely to introduce misleading connotations deriving from other parts of the world. But the general perception of enduring interdependency between host and incomer, and the expectation of gratitude, loyalty, and obedience from the latter as a part of it, are real enough. The feeling of obligation, at least for any but the most abusive, is expected to last a lifetime.[5] Famine and the current era of HIV and AIDS have tested the limits of kinship and hospitality—and surely of gratitude too. Even where orphans have found shelter and clothing, feeding programs are being called for from development and relief agencies just to give them breakfast.

All these questions of obligation surrounding child care and fosterage become important when one considers the recurrent pattern of arrears and defaults on institutional loans, including "farm" loans, perceived as coming from the government and from whites and people overseas: from what Swahili-speakers class as *ulaya*, Alienland. It is not just that there is no concept of moral wrongdoing (in DhoLuo, *rach*, bad or evil) in borrowing and not repaying. It is rather that resources borrowed from distant strangers (let alone from ones that may represent foreign races or faiths, or institutions that represent a mix or none at all) do not carry the same kind or degree of moral claim—or practical one either—as what one owes to one's kith and kin, who may well be needier anyway. Robbing Peter to pay Paul may not be entirely moral, but it beats simply robbing Peter; and a sin of omission (if so it may be called) is not the same as one of commission. All things equal, Luo in my acquaintance would rather have cleared their debts, including those to cooperatives, banks, aid agencies, and other faceless or possibly hostile bureaus. But they usually have sick mothers, children needing school fees, or fathers-in-law clamoring for bridewealth cows too. All things never are equal.

The teaching of morality has in the past fallen partly to grandmothers. These, by their nurseries (*siwindhe*) for children in their homesteads and neighborhoods, have provided a place to sleep with other children, a temporary respite from parental oversight, and a fund of stories, songs, and proverbs that can be a practical and moral education in themselves. That these women have typically married in from other parts of Luo country, on the virilocal pattern, helps spread wisdom and knowledge around and helps homogenize Luo culture to a degree. As Cohen and Atieno Odhiambo have observed, "the pim's (grandmother's) nurturing was the crucible of Luo culture in the past"—that is, before school and church partly preempted her role, and until increasing numbers of Luo began moving into towns (1989, 95). How well a child respected and obeyed elders determined in turn how much others would trust him or her in turn.[6]

That so much of Luo and western Kenyan child care takes the form of chains of responsibility, with ancestors overseeing adults, and with adults loosely overseeing children who care for younger children, accustoms Luo to the expectation that there is always someone to oversee their actions. Trust, obedience, and learning are all closely connected. To repay a loan is to participate in a coached, and maybe supervised, moral act that keeps a chain of intimate kin intact and contented that one has behaved properly. Later in life, when new financial credit comes to an individual from sources unknown or far away, the borrower with no kin or neighbors helping to oversee the loan may lack the feeling of shared responsibility, even if wishing in principle to repay. The continuity and the intimacy disappear—and so may well the loan for the lender.

It has become clearer than ever, in western Kenya and the broader region, that children are raised not just by their parents.[7] Not all who participate will ever be thanked, let alone repaid. Those who have the power to withhold bridewealth cattle and cash, or land inheritance, will be the ones best supported with remittances or nursing in turn in their dotage.[8] Those whose spirits remain around, after death, potentially to visit infliction or infertility upon their descendants, will be those who receive the prayers and sacrifices. The rest will need to content themselves with being part of a tradition that lives.

Lodging and the Flow of Hospitality

When Luo or other western Kenyans go into towns and cities, often for their first extended sojourns as school leavers in search of work and the bright lights, they typically lodge with friends or relatives from their home areas who have preceded them there, and who can not only provide them with shelter and food but also show them something about survival and getting by in their new setting. Since often these are their elders, their hospitality bears much in common with the child caretaking to which many were accustomed in their rural homes of upbringing. The position of the lodger is also akin to that of a *jadak*, a land client, back home in the countryside: this is indeed among the terms used for a lodger behind his or her back, as are *jalaw nindo* (the one taking a turn sleeping) and *jabuoro* (stopper over or camper): terms implying dependence and subordinate status. On the surface, this can all look like freeloading pure and simple, and outsiders are often struck by the forbearance some hosts show in not openly complaining.

But the urban lodging is part of a process, and a flow, of entrustment and patronage; and to be the temporary beneficiary of it is to take part in a system in which one will eventually, if successful in town, be expected to take on the hosting and dispensing role for others in turn. Usually the arrangements are non-contractual, in that the hosts do not ordinarily fix finite duration on the stays of their lodgers or ask them to pay rent or make substantial material contributions, at least at first. These can be negotiated and accepted, but they are not required of kin, as it is understood that there will be ways to call in favors later, back in the rural home if not in town. But the obligations that fall on the lodgers to accept a subordinate social status in the town home, and eventually, once they are established with their own housing, to provide hospitality and sustenance to others coming along behind may be no less binding as moral commitments—depending, of course, on the degree of success they achieve and the relative needs of their eventual followers. Luo people are particularly known for participating in lodging arrangements in this chainlike fashion.[9] As David Parkin observes in the Kaloleni neighborhood of Eastlands, Nairobi, "The good fortune of one man at a particular time is no reason for indebting a less fortunate one, according to the Luo egalitarian philosophy" (1978, 91), but as usual in Nilotic ethics, the moral dimension seems to be geared to issues of relatedness. Parkin notes that "lodgers who are regarded as related by the head [of household] tend not to pay for their keep, while unrelated ones do" (p. 88). Then again, most Luo lodgers in his study were indeed deemed kin.[10]

Sometimes the dependent lodgers in town work directly for their hosts. These include the children and youth studied by Shirley Buzzard in Kisumu in 1978–79: children as young as eight or nine from rural homes whom their parents send to other wealthier or higher-status homes in town, and who tend to be worked long hours, sometimes sixteen hours a day, for their food, clothing, and a little pay to use or send home. Some of these arrangements would be better described as hired help or sometimes perhaps even indentured servitude than as entrustment. But it is worth noting that whereas 44 percent of the 246 wage-working women she studied had "house girls," only 3.8 percent had "house boys," and that the reason they gave was "since they are not relatives and are thought to be more apt to steal or collaborate with thieves, they are considered less trustworthy."[11] Less, it seems, than either boys who are kin or girls who are not. None had house boys whom they deemed kin: house service was a sort of work deemed "inappropriate" to ask of male kin.

Sending children away to school can be called an entrustment in itself, but it is one that leads beyond my competence to discuss in detail. It can of course entail great rewards both tangible and intangible, as high levels of

Luo academic achievement certainly attest. But it has also, in recent years, been a setup for letdown as graduates have had a hard time finding work to suit their learning. Schooling also involves risk of betrayal in more personal forms. In the Kagogo community where I stayed, children in the local primary school, not an unusual one, were made to do farm work on a garden of whose produce they received nothing. It is not uncommon in rural Kenya, or other parts of East Africa, for schoolchildren of either sex to have to cope with physical punishment (which their parents or guardians may or may not practice at home or approve for discipline), or for schoolgirls, at whoever's initiative, to be impregnated or passed sexually transmitted diseases by their teachers or headmasters.[12] Everyone knows secondary school children who have come home from school pregnant, and it is not unheard of in primary either. It happens just as house girls and young women house servants in towns are sometimes wooed, beaten, or made pregnant or sick by the men of their host homes.[13] In the larger picture, one of the benefits—and drawbacks too—of schooling is that it interrupts cycles of serial child-to-child nursing described above, with all the practical work and the work-as-play this nursing involves. But it does not remove them from obligations to elder children, since those who pay their school fees can claim a moral ascendancy over them that can last indefinitely.

As Islam gains a foothold in the western Kenyan countryside, more parents can be expected to send their children to Qur'anic masters for the kinds of group tutelage—and apprenticeship for life—by day or in residential boarding arrangements, that form disciples all around the Islamic world. These arrangements, when they work out satisfactorily, commonly do involve feelings of lasting indebtedness, on the part of not just the disciples but also their families, who may indicate their appreciation by making substantial gifts to the masters, for instance on celebrating completion of the first memorization of the Qur'an.[14] In my experience in other parts of Africa, Muslims speak of this kind of obligation to Qur'anic masters as eternal and as not diminishing, but augmenting over a lifetime and beyond.[15] Another important feature of Islamic residential training is the convention that older children and children better versed in Islamic text, faith, and tradition are often expected to help oversee and discipline (in all senses) the ones who are less so in their circles. In this last sense, at least, Islamic custom is likely to fit closely with local Luo and western Kenyan ways of raising children with the help of jopidi. Where elderly masters are concerned—and they earn their reputations and followings over time—Islamic convention can also fit preexisting

East African patterns of warmth and cooperation between grandparents and grandchildren.

Where children and their shared care and oversight are concerned, the melting distinctions between the economic and non-economic, and between work, play, and learning, all suggest the need for a new or expanded vocabulary of human entrustment: one sensitive at once to demography, politics, and pedagogy. What some call sitting, others call guarding or mentoring; what some call exploitation, others call education or apprenticeship. People who entrust people to people, and in the process teach entrustment itself, are engaged in a flexible, adaptive, and recursive process of nurturance that as yet has few words to define it.

No one knows how old all these patterns of child-to-child nursing, fosterage, and apprenticeship are, or just how long the chains of support and responsibility may stretch through time or space. Obviously they are not perfect: one who looks can still find plenty of abuse and neglect of children and the elderly, child malnutrition, sexual promiscuity and adolescent pregnancy, and dropped moral commitments to distant brothers needing lodging or bridewealth or to out-married sisters needing a place to offload children.[16] A newly emergent and sobering literature on "street children" in East African towns and cities (Kilbride et al. 2000, Flynn 2005) also makes clear how easy it can be to fall through the cracks, or to have to run away, and to have to reconstitute family and community from scratch. Friendship, that product of seemingly unpredictable interpersonal chemistry, here emerges as a critical strategy and recourse, as yet scantly studied.[17]

Certainly there is much flux occurring in the broad picture of child nursing, fostering, lodging, apprenticeship, and residential discipleship involving children and youth in the Luo country and East Africa. If in the past, it could be said that "it takes a village to raise a child," it has lately taken a lot more, including contacts in town for hard times, and the children may come out better or worse off for being handed around. Much research remains to be done on these topics—particularly in an era of HIV/AIDS, disappearing parents, and increased reliance and pressure on grandparents, child nurses, and foster caregivers in town and country.

But the basic principles that have guided the entrustment of children in the past are likely to continue to guide it in the future. Elders are expected to take responsibility for juniors. They usually learn this responsibility very young, by observing and by hands-on practice. They are allowed to command any juniors in reach, and they are accustomed to expecting obedience and respect. They may later lay claim to some of the work, the learning, or the earning of those they nurtured along in youth, paid school fees for, or housed in town; but there are few guarantees, and one may need to be content to be part of a longer-term, serial process. A good Luo exercises the power and privilege of oversight with restraint. Overseen in turn by ancestors if not by elders, one passes on the benefits received from those who have gone before, both kin and non-kin, to those following along behind, even if reshaping them in the process. Tradition passes not just from parents to children, or from schoolteachers to students. It passes too from grandparents to grandchildren, and not least important, from children to children themselves.

Entrustment of Animals

A walker in the Luo countryside sees cattle, sheep, and goats on every hillside, or in the evening, in nearly every rural homestead. But seeing the animals and who is herding them gives little indication of who owns them. Indeed, no one has ever enumerated livestock ownership reliably in or around the Luo country.

One reason the task is so hard is a basic understanding in Luo and East African etiquette that one does not count humans or animals—at least not aloud. Asked why not, some point to old fears it would mean sizing up a target for a theft or raid—as it still might. What they might not articulate is that this sort of counting entails a reduction of quality to quantity that is deemed unnatural.[18]

Another reason why it is hard to find out who owns which animals is the common practice of long-term livestock lending or "fostering" between persons who may live miles apart. (Luo use the term *riembo,* vt., lit. to drive or drive away.) Stock owners entrust out animals commonly to kin or affines, but sometimes to other trusted friends. A "stock associateship," as some observers have called the resulting relationship, can be a special, semisacred bond. Although it requires no special ritual in itself, it is useful for political alliances in rather the way that blood brotherhood once was in much of Africa, or in the way that godparenthood is in other societies where practiced. While a loan defines a stock associate, it is not just a loan; it is a deeper sort of entrustment. Luo are not alone in this understanding: indeed, the stock associateship plays an important role in many of the major East African cattle-keeping societies.[19]

While a survey cannot do justice to the meanings involved in stock loans, it may give some idea of the incidence. Often one needs to ask the questions several ways. What makes the animal exchanges hard to study among Nilotic people is precisely their deep cultural importance: cattle are so bound up with identity (for elder men in particular), and with family wealth and position, that anything to do with them involves delicate issues of self-esteem. In 1981–82, when asked about livestock transfers, about 25 percent of 107 respondents in our Kanyamkago subsample reported that they or others in their homesteads had transferred livestock to members of other homesteads within a year before being interviewed. Most of these animals had been mature cattle transferred singly, but some homesteads had more than one partnership with others. Among the small number of givers and lenders, 51 percent had sent the animals to people they deemed kin (about 63 percent of the recipients being within the same sublocation), and about 71 percent expected their eventual return. Meanwhile, about 31 percent of the total had *received* stock, 84 percent of these from within their sublocation (roughly, clan territory), and 70 percent of them from kin; 80 percent of the receivers expected to have to return the animals.[20]

Another survey conducted among Luo-Suba not far away by Steven Johnson, near Migori town, revealed that while only a minority of homesteads had lent out cattle, most of those that did had done so widely enough as to have multiple claims on animals in others' possession.[21]

Lenders lend for many reasons. Predictably, survey respondents tend to concentrate on the agronomic and economic reasons: to take advantage of more open grazing (the most common answer in our survey), to conserve land in heavily cropped areas, or to slough off the labor of herding where hands are few. But a big underlying reason, not as often reported in such interviews, is that lending one's cattle around spreads their risks of disease, theft, or raiding.[22] As in delayed marriage payments, many use loans to conceal how much livestock they hold: playing poor helps prevent jealous accusations of stock theft or witchcraft and cut down claims for kinship charity or demands for political contributions or bribes.[23]

Borrowers, for their part, accept the animals for the milk, the traction, the benefit of added manure on their fields, and the prestige of a large herd. Both borrowers and lenders value the special trusting bond a stock loan marks. Perhaps the resulting interbreeding freshens herd gene pools. Also, by controlling bush growth, borrowed livestock in less densely settled areas may help control the tsetse fly that spreads trypanosomiasis (sleeping sickness) in humans and animals—it needs leafy, shady areas to breed. Livestock loans are irreducible to individual or collective acts, or to economic, social, or ecological strategies. They can be all these things, and sometimes symbolically loaded too.

A loan or entrustment of animals may last many years, and between close kin, longer than a generation. Animal credits and debts, that is, are both fully heritable. Elders are expected to inform their respective heirs of these before they die.[24] At a Luo funeral, both creditors and debtors of the dead show up to clarify what they owe or expect—a sign of the great importance of credit and debt, and of funerals, in Luo life.

Who keeps the offspring of borrowed or fostered stock depends on the reason for the initial loan. In Kanyamkago, offspring of a cow lent for the lender's convenience will belong, like the milk and manure, to the borrower —a common arrangement.[25] But if the cow is lent for the benefit of the borrower, the offspring are returned to the lender. Some borrowers and lenders agree to divide the offspring equally in alternation.

Some Luo pawn livestock for long periods against other, different stock. The pattern was observed as early as 1909. D. R. Crampton, district commissioner of Kisumu District, recorded that "A man in need of a bull and having himself no male stock would take a heifer to a friend from whom he would receive a bull on the understanding that the first heifer calf born [to that heifer or cow] would become the property of his friend he himself receiving back the heifer originally pawned. As it frequently happened that the heifer in question would drop several bull calves before dropping a heifer calf it will be seen that the animal might be left in pawn for several years. By this time the pawnee would have come to regard the animal and its progeny as his own & would raise considerable opposition to returning it. The most common excuse in this case would be that he had exchanged the original bull outright for the original heifer."[26] Evident here is a kind of "prescriptive right"—that is, a claim accruing by usage and habituation over time.

Livestock gifts and loans are part of sociability. Wealthy men with poorer agnatic kin receive many requests to give or lend cattle or other stock, particularly for marriages. While Luo sometimes speak of livestock in an idiom of individual ownership, and while one cannot be openly forced to share or lend animals, culture influences action with subtle sanctions. The one who refuses may find himself visited by ancestral spirits in dreams, or, if a kinsperson dies unsatisfied, the refuser or one of his progeny may find his or her own plans mysteriously thwarted.

One contributing animals for a junior kinsman's marriage may expect something eventually in return, and cause trouble when it doesn't material-ize. When Odidi Obwonyo of Kanyamkago and his full brother Odhiambo had both died, Odidi Obwonyo's cows passed to Odhiambo's son Nashon for the latter's marriage. Odidi Onwonyo's widow Anyango, who had felt entitled to the livestock, felt hard done by when Nashon, whose marriage eventually succeeded in producing children, gave her nothing in recognition of his in-debtedness to her. She cursed him, and neighbors said they believed that if she died unsatisfied, her spirit would haunt Nashon. Moreover, if Nashon also died, Anyango's spirit might haunt his sons too. (Note how, in this heri-table specter, just as in heritable damage done by bitter money, Luo reveal a concept of surrogacy of accountability in patrifiliation.) Anyango was chal-lenging the male ideals of cattle as exclusive male property and patrilineage property by invoking a rather ambiguous debt. Neighbors construed such a threat as potentially damaging not just to the individual being accused, but to his patriline: so even a challenge to patriliny was to be understood in patri-lineal terms. Stock ownership is qualified; individual and group-based rights overlap. Where rights are shared, so are debts; and where rights are heritable, debts are heritable too.

Debts are transitive in the unwritten algebra of kinship. The sons of a cattle lender may demand that any bridewealth cattle the borrower's descen-dants receive from the borrower's son's daughter's marriage must be handed over until the old debt is cleared. A son who cannot get from his father the cattle he needs to marry may turn rebellious. And there are spiritual con-siderations too: if the son dies unmarried because his father withheld the family herds from him, Luo expect him to return as a malevolent ghost (*jachien*). Most everyone, then, is involved in credits and debts, whether in-curred by oneself or others.

Equivalencies over Time: How Rams Can Become Bulls but Not Ewes

Which animals may be repaid for which, and why? Luo people like to insist that a borrowed cow must be repaid with a cow, and a bull with a bull.[27] In a pinch, however, elders say a bullock or a decrepit old bull will do in return for a strong bull in its prime, and a female calf or aged cow suffices in return for a milk cow in her peak. Switching the ages of animals between loan and return is more acceptable than switching the sexes.[28] The same ap-plies with sheep or goats. The cultural values implied for the different animals exchanged do not necessarily correspond to monetary prices or utility values of the animals, since a young bullock accepted in return for the loan of a bull may in fact be worth *less* in the homestead or marketplace than a full-grown milk cow deemed unacceptable.[29] Local elders' judgments may differ from court judgments, which Luo expect to be based more on monetary equivalencies.

In a sense, rules about sex separation in livestock loans parallel the treatment of gender among the people themselves. Husbands and wives try strikingly hard to keep their financial affairs separate, while fathers feel free to take from the cash earnings of their sons, and daughters are likely to share or exchange food and money with their mothers even when they have moved away from them — as though to live out a perceived continuity of the generations.[30] Luo suggest, in their treatment of livestock and in their own financial lives, that sex is somehow even more fundamental than age or stage in the life cycle.

Sometimes sex differences are deemed even more important than differences of species. One who has borrowed four rams in series from the same lender may return a bullock in their place, Obong' Otieno of Kanyamkago insisted, but he or she may *not* return four *ewes* (nor could one who has borrowed four ewes return four rams).[31] Their prices in the market, though comparable, are not particularly relevant. Culture seems to reinforce boundaries between the sexes where economic equivalencies threaten to break them down.[32]

Livestock in Luoland are a cherished possession, as noted, not just for their productive value and their many culturally specific uses in rituals and ceremonies but also for their role in savings as a multiplying asset not easily divided up or begged away piecemeal. Local livestock loans are not all directly and automatically linkable to institutional credit, and the animals are not deemed properly exchangeable for just any other productive resource. One who asks to borrow another's cow to exchange in turn for chemical fertilizer or pesticides will probably fail. Anyone older is likely to say instead, go get a job.

Some small animals too have strong symbolic associations that impinge upon their exchangeability. A Luo male-headed homestead is normally expected to have a rooster whose presence is symbolically associated with the homestead head's leadership and sexuality, the patrilineage's continuity, or even Luo heritage. This rooster, unlike other roosters, must not be sold.[33] A rooster from the father's homestead plays a key part in the ritual for establishing a son's new homestead. His health and strength are vital for the well-being of the homestead head and the group as a whole; and if he weakens and his comb flops over, he is ritually killed and eaten. If the homestead rooster is sold, the money proceeding from the sale is *pesa makech,* "bitter money," dangerous to its spender and his family. This money and the family must then be ritually purified in an expensive ceremony involving ancestral spirits that Luo consider potentially vengeful. Not only is a homestead rooster not freely exchangeable for a hen, he is not even freely exchangeable for another rooster.

In a symbolic sense, Luo do not consider all livestock freely convertible to other livestock; nor, more broadly, do they necessarily consider livestock freely convertible to other goods. So Luo cultural prescriptions suggest *nested spheres of exchange:* spheres within spheres.

Whether one conceives of such transactions in terms of exchange spheres or in terms of age and gender politics, Luo minds certainly distinguish hierarchies of value separating goods or commodities. Men are disinclined to convert human or animal wealth "downward" into inanimate, contestable, or easily divisible forms, and they try not to do so except in an emergency. Selling animals for grain is a familiar pattern in African food crises. But fertilizer or pesticide is not quite like food. Many Luo would, I think, deem it not just sad but also improper or irrational to lend or sell livestock for such a thing, whether the prices be advantageous at the time or not.

In an economic sense, however, all these things certainly are mutually exchangeable; and a keen observer in a western Kenyan marketplace can find most anything being exchanged for most anything else. There is a slippage between symbolic prescription and economic behavior. Like people anywhere, Luo do not always do what they think their culture prescribes.

Livestock theft and raiding are a complex topic in western Kenya—more complex than can be properly treated here—but a few basic points may be made. The history of stock theft dates back before the earliest written records in the lake basin.[34] It goes on between ethnic groups but also within. It is common in the Luo country to speak of such an act done to one's *own* people with a verb like *kwalo* (v.t., v.i.) or *kuelo* (v.i.), to steal—that is, to commit a sinister, punishable act; and the act done to *strangers* or enemies (which are synonymous) with the word *peyo,* to raid: as a morally neutral act. Kuria, who have been responsible for many Luo cattle losses, have a comparable distinction with their verbs—*iba* and *roona*—only the former being deemed morally sinister because of taking place within a moral community, in their case including any ritually linked clans (Fleischer 2000, 21, 33–34). In a sense, either theft or raiding, if mutual, might be called "negative reciprocity." But people who steal or raid cattle do not necessarily take them from the same people who take from them. They are more likely to take from where defenses are weakest.

Sometimes interethnic raiding, though, takes place with accomplices in the target area. Ties of blood brotherhood in the past have sealed pacts between particular sections of ethnic groups and adjacent ones who would otherwise be hostile—for instance, in the late nineteenth century between Kanyamkago Luo and Isiria Maasai, who also engaged in a "silent trade" of Luo farm produce left in exchange for Maasai animal products without the participants' ever having to come face to face. With their late nineteenth-century pact, they cooperated to pursue thieves and return stolen or raided animals (Butterman 1979, 108–12). Often the participants in raiding are young men who have more brothers than sisters and thus little way to collect from their own families the cattle they will need to pay their own bridewealth. Even where raiding and theft is done by gangs with modern firearms, as it now often is around the Tanzania-Kenya border and in northern Kenya, gaining access to bridewealth cattle and thus to full manhood remains a prime goal (Fleisher 2000). From time to time the Kenyan and Tanzanian governments have both reacted by sending their brutal, marauding paramilitary police, and local communities have also formed their own vigilante groups.

As violent as livestock raiding can be, it does encourage some sorts of cooperation in spite of itself. Fear of raiding encourages animal owners to put out, to "drive," their animals to stock associates in different homesteads near or far, to cut risks of total herd loss. It keeps animals circulating and interbreeding even when things are quiet, reminding people who their friends are.

Farming Out and Folding In

Luo people farm around their animals, I have suggested, by lending and sharing; and they do it much as they farm around their children. They do it for some of the same reasons—demographic, ecological, economic, and social. But the institutions differ too in important ways. Children entrusted to other families may be put in charge of other children there in their turn, and they may be growing aware of not just the power differentials but also the class implications of their shared custody. They are also expected to remember much longer, and ideally their entire lives, their debts for their care, as farm animals can hardly be expected to do. At their best, these customs can benefit everyone concerned by spreading out the mouths to feed and the hands to work, by making sure that both elders and children have care and company, and by exposing children to the norms and values of different families, communities, and generations. At their worst, they can lead to neglect, hunger, and bodily affliction. But they are a necessity and a time-honored tradition no less in eastern and equatorial African contexts; and often have I been impressed, like other sojourners, by the way children and youth rise to responsibility over their juniors when given it.

When it comes to reciprocity in the negative—the raid or the rip-off—it is mainly to cattle and other animals, rather than to children or other humans, that rural-dwelling Luo and their neighbors have directed their attentions throughout recorded history. But it need not always have been so, nor need it always be in the future. Near kin of the Kenya Luo, the Acholi (sometimes called "central" Luo or Lwoo) in northern Uganda, and more distant Dinka, Nuer, and other kin in southern Sudan, have lived through decades of late in which children have not only become the pawns of civil war, kidnapped from one side to another, but actually been among its active perpetrators too (and girl soldiers there have often proven strikingly able and effective at both war and terror). Luo history around the great lake has almost surely involved episodes where children were captured or co-opted from neighboring groups subdued in war, as suggested earlier; and there have certainly been times when children from disadvantaged or lower-status families or larger communities have been assimilated by adoption, renaming, or marriage into militarily or economically more successful ones. These last are indeed among the reasons why the Luo population has grown as large in western Kenya as it has.

At the time of this writing, things are more peaceful. People in the lake basin have been keeping it so even after a legacy of colonial and later state overreach for control, and of secular economic decline in Kenya marked by sporadic episodes of violence and vengeance in city and country. At local scales it is the orderly aspects of child entrustment and nurturing, and of animal fostering too, that are more often evident in the countryside.

CHAPTER 5

Teaming Up
Borrowing, Lending, and Getting By

Wet season debt is settled by dry season debt.
—ACHOLI PROVERB

Getting by in an equatorial African community involves more than a little give and take. But much of the giving and taking is done in expectation of something to be done later: giving something back or somehow passing something along. This chapter describes some of the smaller-scale, shorter-term borrowing and lending that goes on among relatives and neighbors—learned, practiced, and manipulated from childhood on—in the countryside where Luo and other neighboring people live in western Kenya. Human and animal labor, farm tools, food, and money all partake of this fiduciary life.

These things all differ in kind, and each is the focus of its own set of cultural as well as economic values. In an East African understanding, that is, these things are not conceived of merely as commodities, or their circulation merely as markets, since such terms deemphasize the personal relationships and the specific conventions that come into play whenever these things are lent or entrusted. Labor, tools, food, and money can be exchanged for each other on occasion, and East African people sometimes model their dealings in one of these types of things upon their dealings in another. But the personal dimension is ever present. What you borrow and lend, whom you do it with, and how you do it all make up a big part of who you are. This is as true of the shorter-term entrustments described in this chapter as it is of the longer-term ones described in the last.

Some parts of the tropical world—for instance, in India, Pakistan, and Bangladesh—are known for landlords and moneylenders who routinely lend to farmers against the pledge of standing crops, to be repaid after harvest. Some groups on the Kenya and Tanzania coasts, and in coastal West Africa, participate in such traditions. But inland equatorial Africa is different. Rural landlords, professional moneylenders, and tenancy with debt bondage are rarer here. There are differences of wealth and status, to be sure, but coping with poverty here has more to do with relying on a wide variety of neighbors and acquaintances, often in circumstances comparable to one's own. Money lending at interest, let alone high interest, figures less importantly. Luo and their neighbors do make cash loans, but often these take subtle forms that can involve debts lasting many years or even generations. Like loans of land, these entrustments are not made freely or randomly. They rely on culturally specific understandings about the social proximity, or degree of relatedness, of the parties concerned—about who is really responsible to whom.

Local forms of entrustment help understand Luo responses to newer and exogenous financial systems discussed in companion volumes to come. Official lenders attempt to extend credit to unrelated Luo against liens on their crops, livestock, or land. Luo ideas about right and wrong ways of securing loans, discussed in this chapter, are based on cultural values attaching not just to particular types of resources but also to particular kinds of relationships—values that should be taken into account by lenders purporting to inject capital into an economy, build a new mortgage system, or otherwise reform what has not yet been well documented or understood.

Fiduciary Culture—Beyond Finance or Commercial Spirit

As East Africans go, the Luo people are not known for keen interest in commerce and finance. Quite the contrary. Gusii who live near Luoland to the southeast, and Gikuyu across the Rift Valley in central Kenya, scorn the uneven inattention of Luo to their cash crops. Swahili speakers from the coast, and trading people with roots in India and Arabia, all tend to look upon Luo as lackluster traders, earnest enough perhaps but more interested in cows, in clothes, or in big funerals for kin than in their bank accounts, and unlikely to make a go of a business. Luo, these people will tell you, are easy prey for an unscrupulous dealer. Their hearts are not in money. This is the stereotype, and whatever the reasons for it—inland geographic position, disadvantageous climate and soils, colonial discrimination in infrastructural development, or a sharing ethic or other factors attributable to Luo culture itself—there is a grain of truth in it. This is no matter of general ineptness or lax standards—the indisputable Luo achievements in academia, journalism, and dozens of other fields give lie to that. It is more that Luo hearts, at the end of the day, are in other places than their coffers. The manicured Luo cotton or groundnut field, the thriving Luo shop, or the affluent Luo trader or financier is the exception, not the rule.

But to infer from this that Luo have no interest in fiduciary matters would be quite wrong. This chapter seeks to show something of the keen interest that Luo do take in entrustment and obligation of various resources, and to show what these mean for their culture and society, and for familial and personal identity. If, I suggest, the Luo people take as keen an interest in loans and debts as they do—despite their evident general disinterest in commerce and finance—then perhaps other people across rural East Africa do too. Perhaps for them, as for Luo, fiduciary culture is not just about money, or about profit.

Many things in Luoland are lent or entrusted around, and for many reasons. Since land, animals, tools, labor, food, and even children all move around in exchanges expected to be reciprocated after time, to enter the Luo country with an idea of loans and debts as being about money is to miss most of the action. People borrow and lend, entrust and repay, for many sorts of reasons: to share, to spread out risks, to hide their wealth or shelter it from third-party claimants, to speculate on seasonal or longer-term price fluctuations, to take advantage of local imbalances in resource endowments, to establish or clarify status hierarchies, to gain voters or supporters, and to create or maintain social relationships useful or satisfying in themselves. The loans and entrustments that Luo make between themselves, and among intimate neighbors, may last only minutes, or as long as several generations.

These things are easy to misunderstand. The tendency of international development agencies since the second world war to gravitate toward credit as the main way to approach rural people has been based on an assumption that loans are otherwise hard to obtain on reasonable terms (and that rural people are too poor to save). "Development" project documents tend to treat institutional credit as an essential input in agriculture, like soil or sunlight. Once institutional loans are issued, lenders disappointed in repayment rates —as they have nearly always been—usually explain them in terms of technical problems in their schemes, local inexperience to credit, or what they conventionally call "moral hazard" on the part of the borrowers.

While specialized rural moneylenders are scarce or absent in rural Luoland and elsewhere in the Kenyan countryside, it is wrong to infer, as foreign economic analysts have sometimes done, that farmers of the area have no access to credit, or that credit is a new idea for Luo.[1] As I describe in the following pages, Luo speakers deal in a great variety of loans and entrustments, some monetary but most not. These, and the debts and obligations they entail, pervade Luo life—and even involve the Luo world of the dead. A term like *credit*, superficially connoting bankers and signed contracts, scarcely does justice to all the kinds of borrowing and lending one finds in a rural African setting. Nor are these any more explicable as economic than as cultural or political phenomena.

Without a broader understanding of Luo credits and debts, the outcomes of "formal" lending programs make little sense. While not all strictly economic, the diverse debts and obligations that Luo carry help explain why loans from institutions like credit banks and cooperatives often go unrepaid. The problem here is *not* that Luo are unfamiliar with credit. It is rather that foreign creditors are too unfamiliar with *Luo* credit to understand that their institutions will be low on borrowers' pecking order for repayment and other cooperation.

Most everyone in Luoland has debts already, and obligations to collect upon, before any bank or cooperative lending institution enters the picture. Webs of indebtedness stretch across the countryside, connecting into cities too. Rural people have their own informal hierarchies of creditors; and the latest, most distant, and least familiar lenders usually rank at about the bottom.

The local loans and obligations are not just dying vestiges or quaint archaisms but often vital, adaptable ideas and practices; nor is their interest purely academic. They can help show, for Luoland and elsewhere, whether credit is an appropriate form of "development" assistance or relief in the first place—something I suggest not be assumed—and if so, what kinds and for whom. For what purposes do Luo people have trouble borrowing locally, and why? Should institutions lend resources without having taken savings deposits? Should they lend to groups or individuals? In cash or kind? What makes useful security? Under what circumstances should a loan be written off? Indigenous ways of borrowing and lending might themselves be understood as a fund of ideas from which outsiders might borrow in fitting financial interventions better to African needs and aspirations. For more are relevant than usually supposed, and their implications are profound.

We now continue with our brief survey of Luo borrowing and lending, and of the entrustment and obligation within, beneath, and beyond. Underlying all that follows, of course, is the possibility of borrowing and lending land. This big topic is treated in a volume to come. But it should be borne in mind that all of what follows can be conducted on land where one is considered to belong as a native (that is, by someone living among kin and ancestral graves); as a stranger tolerated at the sufferance of natives and neighbors; or as one betwixt and between, which is the position in which women find themselves upon marrying into their husbands' lineages and homes. The fiduciary culture of landholding, like much of what is to follow, ties together the sacred with the secular, and the symbolic with the economic and political.

Labor Exchange, Share Contracting, and Their Social Calculus

Even with the efforts of children and animals, labor for farming in Africa can be scarce. Some think it is the tightest bottleneck. In the continent with the sparsest overall population densities, the one with hand hoes most widespread as a main means of tillage, and the one where farm work falls most heavily on female shoulders, the conclusion is widely defensible. Population growth is swinging the arrow toward land as the main constraint, particularly in the more heavily settled areas like the Luo country and the still more crowded Gusii and southern Luhya areas adjacent.[2] But even there one finds seasonal and longer-term labor constraints.

Over the annual cycle and the family cycle, neighbors cope with the crunches by give-and-take, by borrow-and-lend, and by hire for cash or kind. To give and give without somehow taking is hardly human nature. And Luo do not particularly like to engage in wage work for neighbors or kin, though it does happen—it compromises that delicate self-image as sharers and co-operators. The peak need for weeding, the longest task, comes during the lean season, which is also the wet, malarial season; and western Kenyans often lack the ready means to compensate help on the spot. So as often as not, it is borrowing and lending that smooth out the peak demands. It is a sociable solution—much of it goes on in groups. It is not, however, always a wholly altruistic one, or as egalitarian as it may seem.

Work's definition is culturally specific, and the word does not translate neatly between all tongues. Luo country people do not, for instance, usually call reading and writing *tich*, work. Often, after many hours transcribing interview tapes or writing up my notes, I have had a neighbor come by, peer through the door at me at my desk table, and say, to be polite, "Iyueyoyueya": so you're just resting, resting. Tich, real work, is physical work, or a recognizable job. East African cultures vary quite a bit, too, in task allocation by gender and age. The Luo gender division, as I have observed it, is complex: in farming, herding, house building, and so on, males and females tend to work on tasks that are separate but interdependent. The age division is much less sharp, as children learn and practice early in life many of the tasks they will perform later on.[3]

Luo people exchange labor for both farming and herding.[4] They also exchange it for the child care that can seriously constrain women's and girls' farm work. Since people rarely work each others' fields or herds in reciprocal fashion simultaneously, many mutual exchanges may be understood, by insiders or outsiders, as entrustments or loans of sorts. Exchange labor takes several forms, each familiar from other parts of the continent. Simplest are dyadic partnerships within or between homesteads. Most of these follow kin ties, and most often men work with men, and women with women.

Luo neighbors borrow, lend, and combine personal and family resources in countless individually tailored arrangements for farming, many involving loans and counter-loans arranged ad hoc for a season only or altered from one season to the next. These are not all "share contracts," the arrangements in which parties contributing resources agree ahead of time to divide output in prearranged proportions; some are arrangements whereby farmers harvest their own fields for themselves. But like sharecrop ties, they are not just economic *contracts:* they are social *relationships* too.[5] Exchange of goods or services, and obligations incurred thereby, are essential parts of most any social relationship.[6] Land for labor, labor for an ox, an ox for some labor and the use of a plow — loans and counter-loans are central but variegated threads in the fabric of Luo neighborhood life. What you lend or borrow helps define who you are. An able-bodied man who lends out livestock against neighbors' labor is likely to be deemed a person of substance (*japith* or *jamoko*), while one who begs ten shillings' worth of maize, promising to work it off over two days, will be deemed a pauper (*jachan, ng'a modhier*). These things are relative; by borrowing or lending, one can become a patron or client to the loan partner, and one can be both at once with different neighbors.

When sharing work in neighborhoods, many Luo prefer to be compensated in food rather than cash, since cash defines one as a laborer (*jatich* — the term also means "servant") and suggests finite commitment on the part of the hirer.[7] Luo like to say they do not engage in wage work for one another in the countryside. In reality, however, it happens often, and sometimes even between full brothers.[8] Labor loans shade into hirings but also, alternatively, into gifts or favors.

Short-term loans of draft animals and plows between members of neighboring homesteads are an important mechanism for coping with poverty and uneven resource distribution.[9] While only 67 percent of the 107 homesteads in my Kanyamkago subsample possessed ox plows, and about 77 percent possessed draft animals, 95 percent prepared their land with plows in the 1982 long rains season. About half of those without plows or draft teams borrowed them, others renting them or hiring neighbors to plow for them. Among Luo, as among nearby Suba who now speak Luo, a homestead in possession of one draft animal commonly teams it up with the animal of another.[10]

Partnerships formed with different contributions are instructive. Where two farmers team up such that each contributes one ox, but only one of them contributes a plow, he or she does so without extra compensation—both homesteads use the team for equal time. In a second kind of exchange, an ox owner lends it to a plow owner, and the latter lends him the plow in return, again for equal time. In a third version, an ox team owner and a plow owner ally to rent out the oxen and plow to a third party, and they split the earnings evenly. Now, a good ox in South Nyanza in September 1981 could be bought for about Ksh. 800 (U.S. $96) in the marketplace (the exact price depending on time, location, and quality, as well as buyers' and sellers' identities and circumstances), and a new plow with accessories, about Ksh. 400–450. But Luo were prepared to neglect such differences in construing an exchange *as if symmetrical.* While a plow was treated in the first context as a negligible contribution, in the second it was treated as the equivalent of one ox, and in the third, it was treated as the equivalent of two oxen (which together would fetch up to four times as much in the marketplace). In each kind of exchange the partners behaved *as though their contributions were equal.* What was important was not the relative cash values of the goods put in but the fact that each party was contributing something essential. Up to a point, it seems to be worth more for a Luo to be a member of an egalitarian partnership with hope of continuation, and to keep resources moving, than to get the maximum economic return from a specific contribution.[11]

Plow team owners sometimes lend their teams free of charge to close kin or friends. As Alloyce Onyango of Kanyamkago put it, their decision to do so depends on their *chuny* (lit. liver; or in English, heart, spirit, or disposition). It also depends on who is concerned. Four days at most are proper, he said, except for a daughter or daughter's husband: they can have up to three weeks. Perhaps this is because the affinal tie is so delicate, and so vital to reproduction as well as to animal circulation. In ideals at least, chuny ties economy closely to kinship.

But lenders can gain too. Loans to distant kin or unrelated neighbors help win votes, courtroom support, or other favors. Commonly the borrower agrees to plow the lender's land with the team before seeing to his own fields. These livestock-for-labor arrangements are asymmetrical in power and patronage; Luo, like other western Kenyans, deem the labor provider subordinate, and office holders sometimes dispense the favors strategically.[12] I have observed, however, that the durations of the loan and counter-loan can be adjusted to absorb the plow driver's economic debt.

Some farmers whose landholdings exceed their immediate needs lend fields to neighbors, for a season or a year, in return for the use of plow teams: an unwritten share contract sometimes called *bar-wabar* or *bar-wa-bar* ("we divide-divide," from *baro*, vt., to divide, split, or chop).[13] Usually borrowers and lenders of plow teams agree on a fixed loan period, such as four days. Most plowing is done in the cool hours of the morning, and some farmers who borrow plow teams for fixed periods like to sneak out to their fields in the small hours to drive the animals longer without being seen to overwork them. They thus try to gain the most from their loans while keeping the lenders open as sources of credit in the future.

Occasionally Luo devise triangular exchanges for plowing. One such instance, in Kagan, went as follows. Henricus Okech and a neighbor, Meshack Mwalo, both owned ox teams and plows. Philip Ojwang, another neighbor, had none. Ojwang and his family did the work of plowing for Okech and subsequently took the team and plowed his own fields. His fields were larger than those of Okech, and he had to spend more time on them. He thus incurred a debt to Okech. He had little cash with which to compensate Okech for the difference. Mwalo agreed to pay Okech what Ojwang owed, and in return for Mwalo's financial help, Ojwang plowed his fields for him for the season. In this case Ojwang could be considered a client of two patrons. Like many others, this triangular exchange is an unnamed occurrence, an ad hoc invention of its parties.

Luo bend their borrowing and lending methods flexibly. They adjust them not just to accommodate economic differences between individuals or homesteads but also to accommodate their idea of local custom. They have devised re-lending arrangements to cope with problems in "formal" credit schemes: farmers whose government cropping loans delay in the planting season borrow oxen, plows, labor, or cash from their neighbors, using the government's loan approval as a guarantee of ability to repay. (And the triangular brokerage pattern recurs here too.) Cash is finding its way into loans and other exchanges of land, livestock, labor, and other productive inputs, but the transition to cash does not appear steady or inexorable.[14] Other forms of compensation remain attractive to farmers and are certain to persist in the future, instead of cash or in conjunction with it.

Whether share contracting and related exchanges altogether benefit their parties mutually in the end by evening out imbalances, or boost production for some while diminishing it for others less fortunate — or both — is a matter of current debate I cannot pretend to resolve here. Certainly they are plastic: "Given the inherent instability of the family as a productive unit, and many centuries of erratic growth worldwide, certain relations of production have taken root because they are flexible and adaptable. The most prevalent example of such arrangements is share contracting" (Robertson 1987, 1). They are not perfect and can sometimes be exploitative. This can depend on whether the one who continues to work as a client for the same patron can eventually gain a more permanent stake in the land or other resources over time, a point on which legal systems and customs seem to vary widely. What is sure is that share contracting will not soon be swept away, by law or by money.

Labor Exchange in Groups: *Rika, Saga,* and *Shirika*

Work in equatorial Africa is something to do with others if you can, alone if you must, and in midday sun only when desperate. Sharing work can be conceived in more than one way. It can be seen in an economic idiom as exchanging labor, in a social and experiential one as keeping company, or in a political one as practicing some sort of solidarity or making a statement about ideology. It is a sign and a seal of religious commitment, since much work on farms is done by church groups, and whether or how often you show up can count you in or out. It is the mix of motives, when these coincide, that make the topic interesting. Government officials and financiers in Africa like to tell rural dwellers what sorts of groups to work in — as families, collectives, and the like — but rural-dwelling people resolutely make up their own minds about this, making their own contacts and finding their own middle paths.[15]

In a common Luo pattern, a small group of kin or neighbors — again, usually single sexed, and often lineage based — forms to exchange mutual help in rotation. This system is termed *rika* (n.), a word with both Nilotic and Bantu variants. Between about three and eight people, most often women, spend their mornings together at a busy time of year, and working alone or perhaps in different groups at other times of day and times of year. The group may endure for more than one growing season. Co-wives commonly form rika groups, but more generally, members of rika groups are a minority. In our Kanyamkago subsample interviewed about labor exchange in 1981– 82, about 58 percent of 107 homesteads had members who had participated in them at some time. But I estimated that only about 15 to 25 percent of women and less of men belonged to such groups currently, and members of inter-homestead rika groups spent only a fraction of their work days over the year in rika labor. Still, the principle of rotating cooperative work is highly valued in Luo culture, and it is adaptable. Some Kanyamkago Luo have re- cently formed rika groups for baling tobacco, newly important as an indus- trial crop. Luo are now forming cash contribution clubs, described below, upon the conscious model of rika; and these too are mainly female.

A related form of collective labor, known as *saga* (n.), involves larger groups of neighboring homesteads that may convene members less fre- quently or only on single instances for work events of a more festive kind. While the work is basically cooperative, the workers often compete to show off their strength or skill — for instance, to hoe the farthest in a set of paral- lel rows.[16] Luo convene saga gatherings for planting, weeding, or other large tasks like smearing a house lattice with mud. These are tasks that can bene- fit much from being performed all at once. The groups can be single sexed or mixed, depending on the task, smearing groups usually being all female and most others predominantly so. The recipient of rika or saga labor pro- vides food and drink, which at least until recent decades included local grain beer and sometimes a slaughtered animal, for the workers for the day. The food and drink are not considered in a strict sense as a payment, but they are just as much an obligatory ritual counter-gesture by the host, who remains indebted to the workers afterward. A first-rate saga, I was told in Kanyam- kago, should include a midday break for porridge, potatoes, or *nyoyo* (a mix of hominy and beans) and end in the sponsor's house with both roasted and stewed meat from a slaughtered ram or at least chicken, and with beer, tea, or other potables — the sexes may separate at this point and be served dif- ferent drinks. Some enduring saga groups are tightly enough organized that any member can convene one, and that the group can exact fines (or, construed otherwise, cash contributions) from members who fail to show up or send relatives as substitutes.

The Luo rika and saga have parallels with similar names among Luhya, Gusii, Kuria, and other neighboring Bantu speakers, and among other neighboring groups including the Kipsigis of the Kalenjin macro-group sometimes classed as "southern Nilotic" speakers.[17] Like the comparable festive labor gatherings that Andeans know by the Quechua term *minga* or *minka*, the large Luo saga is considered by both insiders and outsiders as a "tradition" on the wane, which hangs on mainly in the drier areas downhill by the lake.[18] Mature Luo uphill in Kanyamkago have told me they recalled attending twenty to thirty days of saga work yearly in their youth, confirming Roscoe's impression of 1915 (p. 276) that Luo men and women "often work together in parties on the same field," singing as they hoe. But most of them and their children now attend about ten or fewer a year (house-smearing saga of ten to twenty members remaining common), and some none at all.

One reason for the decline of saga farm labor, if true as locals consistently say — may be that the hastier, sloppier work of a festive work gathering is suboptimal for extracting the highest yields per unit of land, and the losses become more significant as rural crowding shrinks fields over time. Some Luo also attribute the decline to the spread of cash and to what they perceive as rising individualism: "People are too hungry for money now," commented Phillip Ojwaya of Kanyamkago.[19] Of course, on this topic, the past is often viewed with rosy spectacles. Indeed, the appearance of new forms of collective labor — for instance, in tobacco growing — qualifies the claims.

Related in nature to the rika and saga that Luo deem "traditional" are other collective work groups, often church-based, identified by the Swahili noun *shirika*, meaning "sharing, being partners, or acting in common."[20] People who might once have organized weeding and other labor-intensive tasks on a kin basis have lately often turned to a church-member basis instead — especially women, who after marriage usually live among their husbands' natal kin rather than their own. As the foreign term suggests, Luo consider the shirika a more recent and "modern," post-Independence (1960s on) kind of group, whatever its real age, probably older. Not all members of a church, which may number several hundred in a congregation, attend its shirika; these gatherings are much smaller. Single sexed or mixed, a shirika group commonly meets once weekly — the seven-day week being a post-European contact mode of organizing time — and members of the work group who fail to show up typically pay a cash fine of about a day's casual work wage. Churches often send their shirika groups to help weed for new members, partly as a way of attracting new recruits, or hire out their services, offering members a discount. Some among the Luo have treasurers who collect funds to lend or donate to members with special cash needs, such as to make up school fee shortfalls or to transport a body home for burial. Some have hierarchical offices and have been registered with the government as eligible to receive loans as self-help groups, and some have opened joint savings accounts in commercial banks. In sum, the church basis, the weekly schedule, the cash, and the bureaucracy where involved all give the shirika the stamp of modernity.

While Luo and some other western Kenyans value collective labor and the egalitarian ideals of cooperation that attach to it, it remains a status marker and divider. Wealthier Luo engage in saga farm work far less often than poorer ones, often preferring to contribute cash for the food buying. Early in my stay among the Luhya of Isukha, I asked a relatively affluent woman, whose husband was out supervising an expensive rented tractor, whether she participated in the neighborhood *lisanga*, a women's work group. She laughed and exchanged a furtive glance with the young local man interviewing with me, as if to say, "what a ridiculous question!" For their own farms, wealthier Luo or Luhya may hire such groups, but some prefer to hire individual laborers whose services they think less likely to entangle them in multiple obligations. For poorer farmers, collective work gatherings of almost any kind remain an important way to obtain a substantial meal in hard times, and those multiple obligations can be a lifeline.

So far, this chapter has suggested that many kinds of labor and livestock exchange look like loans, or if not called loans, rely heavily upon entrustment in ways culturally prescribed. Loans variously merge at their edges with gifts, sales, or even thefts, but they do tend to follow conventions. Individuals tinker constantly with entrustments, and aggregates demonstrably adapt new kinds, but they often base these on older conceptual models, as when they base tobacco labor groups, or money contribution clubs, on older principles of rotating labor. Some loans involve partial counter-transfers on the spot, as if to cancel out obligation; some involve reciprocity with obfuscated imbalance. In important ways, whom you work with, and on what terms, determines who you are. Below we begin to consider some longer-term loans and entrustments involving not just humans' and animals' services but their basic sustenance.

Relative Obligation: Food Loans to Get By

Food loans, within and between homesteads, seem to help bind neighbors together and cut down both local and regional inequalities in farm production. For domestic use, Luo people most often borrow food from April to July, as granaries run low, and in quantities of less than half a ninety-kilogram sack. Women do most of the arranging and conducting for small food loans.[21] Larger loans and entrustments—for instance, sacks of grain lent to neighbors for use in funerals in other homesteads—are more often effected as transactions between men, usually ones deemed homestead heads (*weg dala*).[22] Bigger as well as smaller transfers of food, more often than not, involve perceived kin ties of some kind.

People who lend or make gifts of food between localities often follow matrilateral or affinal bonds. Here the rule of clan exogamy sets up convenient exchange partnerships. These are especially useful to lowlanders, whose farming tends to be riskier, and who often rely on seasonal food gifts or loans from uplanders. Daughters once married away (since virilocality, residence at the man's place after marriage, remains the norm) usually bring their mothers food or other gifts when visiting. These are sometimes construed as discharging obligations incurred at the natal home earlier in life. My information suggests that between close kin like siblings or agnatic first cousins, food is likely to be given freely, or repayment postponed indefinitely; more distantly related kin more usually expect repayment just after the harvest.

Luo people lend food between homesteads with no explicit expectation of "interest" payments, though it is deemed polite to return a slightly larger quantity, rather as it is expected that a maize flour seller in the market will add a small scoopful of grain to the amount agreed upon to confirm a relationship.[23] In the case of a pre-harvest to post-harvest loan, an equal quantity of food returned represents *negative* interest in economic terms, since the exchange value of the good declines upon harvest. Seasonality determines just how much of a favor a loan is.

Where a borrower's harvest fails completely, he or she feels no compunction to return the loan—this is a risk of the lender. Luo people carry expectations of interest-free loans and of crop-failure "insurance" over to institutional credit, and these are sources of misunderstandings between them and foreign creditors.

Just as a plow-team lender is locally deemed superior in a sense to his borrower, the principle applies too to loans of most other things. Lenders boast of the help they have given needier kin or neighbors. (It is always much easier for the researcher to learn about items lent or given away than about items borrowed or received freely.) The pride, power, and prestige of the lender are important elements of long-term as well as short-term credit.

Kinship seldom governs social behavior by itself—a point on which many ethnographers in East Africa have presumed too much. But perceived relatedness does carry expectations, and kinship is sometimes adjusted to square with real behavior. As one way of sorting out what kinds of gifts, loans, and sharing male Luo *expect* between particular kinds of kin, a series of interviews were conducted with three men from Kanyamkago, of whom two had already been trained to read and write genealogical diagrams.[24] With reference to a given individual (or "ego") on the charts, they were questioned systematically about relationships with kin. We gradually spread outward laterally (that is, toward distant cousins), and vertically (toward distant ancestors and descendants), taking relationships of both consanguinity ("blood" ties) and affinity (marriage ties) into consideration at each step. With each kin bond, transfers of several possible things requested were contemplated: an acre of land, a plow team and plow, and a sack of maize in a hungry season when the person approached has a small surplus. Hypothetical questioning is not, of course, the only approach—it is important to collect real cases too, as we have attempted in surveys and (still incomplete) case studies, since ideal and real behavior often do not match. But the more abstract approach helps separate ideas about kinship from individual personalities, friendships, and other circumstances.[25] Here are summarized, then, a few basic cultural expectations about men's maize loans that the interviews revealed.

> *Key to abbreviations:* following anthropological convention, F = father, M = mother, B = brother, Z = sister, H = husband, W = wife, S = son, D = daughter. Compounds formed thus: ZHB = Sister's husband's brother, MBS = mother's brother's son, and so on.
>
> 1. *Kinship means privilege.* A kinsman, even if related only through a series of consanguineal and affinal ties (for instance, a FZH, MZH, or even a BWBS or DSWB), should be given or lent maize automatically upon request. But an unrelated man should not.
>
> 2. *Patrilineal kinship is privileged over other kinship.* To a F, FF, FFF, or a S, SS, or SSS, a man should not *lend* a sack of maize but give it freely when asked. Most other male kin under the same circumstances should receive loans or, if the maize is given freely, the amounts should be reduced. The male line takes precedence over the female. If a FFF and his wife should receive a free sackful, a MMM and her husband should get only a half sackful. Similarly, if a SSS

and his wife should get a free sackful, a DDD and her husband should get only a half sack.

3. *Affines related through male consanguines (from a provider's perspective) are privileged over affines related through female consanguines.* If a BWB should receive a sack of maize, a ZHB should get only a half sack.[26]

4. *Affinal ties weaken in fewer steps than consanguineal ties.* If a BWB (being related through one marriage and two siblingships) should receive a full sack of maize freely, a WZH (being related through two marriages and one siblingship) should get only two debe tins—that is, less than half a sackful.

5. *Expected economic transfers do not always correspond with kin terminology.* If a WF receives a half sack of maize, a WMZH should receive only half a tin (that is, a tenth of a sack) even though both are referred to by the term *jaduong'* (father-in-law or classificatory father-in-law; also "big one" or elder). Similarly, if a SS receives a full sack of maize, a paternal half-brother's son's son (FSSS) should receive a half sack, even though each is called a *nyakwaro* (grandchild or classificatory grandchild) and though both are of the same generation.

6. *The very old and (to a lesser extent) the very young merit special attention.* An unrelated, unacquainted person within two generations of one's own (whether ascending or descending) should be given nothing freely. But someone of one's own great-grandparents' generation should be given a sack of maize freely, because, informants explained, he must be old and obviously desperate, and because he might otherwise curse one or bring spirit vengeance if refused. A non-kin of one's great-grandchildrens' generation should be given a tin (less than a quarter sack) of maize freely; he is likely to need the help because of his youth and inexperience.

Might these cultural priorities be somehow adaptive? A sociobiologist might construe the several principles favoring closer kin as subtly protecting one's own genes in the gene pool, and thus as promoting "inclusive fitness." But data stretching over generations are too scarce to prove the theory at present. Patrilateral biases are especially hard to explain ecologically. The principle favoring the very old and young may well serve to buffer the weak and vulnerable against malnutrition or famine. But to explain elder-protection in terms of adaptability is harder, since they may well be past reproductive age. To do so requires seeing their cultural usefulness as repositories of wisdom or experience: for instance, knowledge of plants edible in emergencies. Ecological explanations will be weak without culture woven in.

Money Lending without Moneylenders

Farmers in western Kenya borrow money mainly from people who do not specialize in lending: they borrow from shopkeepers, other merchants, and the larger farmers. Specialized moneylenders seem most commonly to be found in either societies characterized by highly uneven distribution of land rights and by rent tenancy (as in some South Asian societies where landlords themselves often lend to tenants, sometimes using their superior land rights to keep the latter in debt bondage), or societies where a class or ethnic group is denied access to productive land altogether (as with Jews in nineteenth-century Poland). Perhaps, if landholdings become concentrated in fewer hands in western Kenya, money lending and moneylenders will gain importance. If so, they will likely be a symptom, a short-term remedy, and a contributing cause of land poverty all at the same time.

As it is, local cash loans for farm inputs are hard to come by, even between friends and relatives. Uterine brothers who will freely lend each other livestock for plowing sometimes refuse to lend each other cash for the same purpose, mainly because lenders find cash harder to recover. (No foreigner who has lived in Luoland will have failed to notice how much harder it is to recover cash loans from friends or neighbors than loans of durables like lanterns or tools.) Cash may present problems to lenders because of its divisibility and multiple uses, or less often, because of the difficulty of remembering amounts lent. Another possible explanation is that Luo feel morally less bound by loans of cash than by loans of other things, perhaps because cash symbolizes markets and private gain, or the nation-state (whose president appears on notes and coins) and its foreignness.

Neighborly cash loans for farming are small. In a 1965 survey in Kabondo Location, South Nyanza District, 216 farmers were asked a series of three questions supposing that they needed money to buy farm commodities, from small to large: (a) a *panga* (machete) or new hoe, (b) some seed or a plow, and (c) some graded cows or fencing. In each question they were asked whom they could approach to borrow the money. Farmers consistently indicated friends, neighbors, and relatives for the small loans of the first question but increasingly mentioned government offices of several kinds as the amounts in question rose.[27]

Nearly twenty years later, in our study, farmers answered these questions in very similar ways.[28] Interviews and observations suggested that those who succeeded in persuading shopkeepers, grain-millers, or others solvent in cash to lend them small amounts of money for recurring production expenses (for tools, small quantities of seed, and so on) were usually close relatives or friends of theirs. No one appeared to depend on such informal cash loans on a regular yearly basis, though some may have done so without admitting it.

Although farmers find money hard to borrow unofficially for recurrent farming expenses, they can and do borrow for emergencies — for instance, a hunting, hoeing, or wood-chopping accident needing transport and medical attention. But even this can be hard to do, even where there are shopkeepers with cash they could lend. An accident in unusual or sudden circumstances may scare others away from helping out, if only out of suspicion that witchcraft might be at play and the helper might become another target or somehow get tainted by the misfortune. Such a concern may be unspoken about (and some Luo and Kenyans say speaking of witchcraft invites it or accusation of it), but such concerns may help account for cases where there seems to be a dire need and onlookers but no one around willing to act.

Sometimes the help comes from a group. Some Luo raise large sums from lineage-based or locational welfare groups to avert distraint of mortgaged land, or to pay hospital fees. Church groups and groups of neighbors who exchange labor (rika groups) are also activated for personal emergency loans. These associations, like townspeople's larger civic lodges in western Kenya, are often multi-purposed; and farmers sometimes turn to them as small-scale financial institutions in financial emergencies.

Rotating saving and credit associations, also known as contribution clubs or known by other names, are a form of self-help association most common among small-scale merchants and salaried people in and around towns. At the most basic, they work like this. Members meet at regular intervals, each chipping in a fixed amount. One member takes home the pool at each meeting, until everyone has had a turn, whereupon the cycle may recommence. Rotating saving and credit associations are also sometimes formed or joined by farmers, but they tend to depend on steadier contributions than most people who depend mainly on rain-watered farming can muster. The associations are a way friends or neighbors can gain each other's help in saving up for some of the larger household purchases like sheet iron roofing, or for stocking retail shops. They are also used as places to turn for emergency help. They are handy ways of saving and borrowing, even and especially without official help.

These clubs, like the habit of "overselling" crops at harvest and buying back later, are a way of handling money designed with social relationships and impressions firmly in mind. For not only are they typically based on pre-existing ties such as neighborhood, sports club membership, common trade, or the like, but they serve as a way of accumulating capital without seeming selfish to other needy kin or neighbors. Every contribution made is on other members' behalf as well as one's own, and every withdrawal may be monitored by members who will help ensure that it is used in whatever way the individual and group have planned. The member may thus be able to get away with making a fancy cash purchase or investment without being so likely to incur malicious gossip or accusations of witchcraft.[29]

Over the longer haul, some kinds of loan or entrustment are expected to be not strictly reciprocal but instead serial in nature. Both urban and rural Luo agree, rightly or wrongly, that formal schooling is the way forward for their children; but beyond the compulsory primary level, it is expensive.[30] A standard solution is that parents, parents' siblings, or other close kin contribute substantial amounts—sometimes full tuition—to a child's or youth's schooling, with the tacit understanding that when he or she leaves school and finds a job, he or she will send something back, if not to the initial helper, then perhaps to that person's dependents, who may need similar patronage themselves for their own schooling. An elder sibling, once out of school, is also expected to help pay school fees for a younger one, thus lightening the financial burden on the parents. A debt to a senior kinsperson, repayable instead to (or for) a junior one, is a characteristically East African pattern. The maxim "wet season debt is settled by dry season debt" can work for credit too, as one generation covers for another. Although Luo and their neighbors tend strongly toward patriliny, not all such assistance follows kin ties through males. Among Luo acquaintances of mine, matrilateral and affinal kin (in-laws) have sometimes also contributed to juniors' school fees.

Obligations like these often take a decade or longer to return or pass on. The patrons usually do not keep strict track of the obligations or hold the junior kin to any particular schedule, quantity of repayment, or "interest." In Marshall Sahlins's terms (1974, 191–204), these arrangements are rather in the nature of "generalized reciprocity." Moreover, the economic favors can be repaid with other kinds—for instance, campaign support, a valuable job introduction, or tutelage. But for a middle-aged market crop grower or wage earner, a promising young kinsman's school fees are likely to represent an investment in his or her own retirement. For people without formal "social security" or pensions—to look through a foreign optic—this can make all the difference in the world.

Luo labor migrants often ride on the contributions from kin who hope for eventual remittances. "Write if you find work" means not just a letter. School fee and migration assistance as long-term quasi-loans are not just Luo practices, of course—I have observed them in other rural African settings as far away as The Gambia. Shorter-term reciprocal exchanges of food and money also flow both ways between urban labor migrants and rural kin, sometimes shifting like tides. In Kenya, quarter sacks of maize—the twenty-five kilogram headload has until lately been the legally allowable limit transportable between districts without a license—piled up atop *matatu* pick-up vans. Money moves by postal money order or in envelopes in the pockets of traveling acquaintances. These flows between city and country kin constitute an important equilibration system in counties where rain-fed agriculture, price fluctuations, fuel shortages, intermittent hoarding, and political instabilities can make food supplies irregular in either kind of setting.[31] Far from being just economic transfers, these actions give substance to ideas of social connectedness.

Securing Loans: Entrustment and Counter-Entrustment

Sometimes, when a lender makes a loan but does not entirely trust the borrower to repay or reciprocate, the loan is made upon the condition of a counter-loan or counter-entrustment, a conditional material transfer, in the other direction. The action goes by many names, including pledging, pawning, hypothecation, and (where deadlined) mortgaging; the thing transferred or conditionally offered against the loan also goes by various names, including pawn, security, collateral, or loan anchor. In DhoLuo, they are mostly covered by *singruok*, a concept also used in discussing religious covenants—a point to which we will return. The question of issuing or pledging loan security (that is, laying down collateral for credit) has become one of the most vexed in Luo life, and the experiments, problems, and misunderstandings surrounding this as a result of intercultural contact are a central subject of another accompanying study. Here only the basics need be noted.

Luo people, and others in equatorial Africa, have until recent times been accustomed to some kinds of lending collateral and not others. Luo have long practiced and accepted the use of animals, eucalyptus and fruit trees, commercial buildings, and tools as objects of loan security—that is, collateral against loans in cash or kind. They use sponsors as witnesses or (if literate) cosignatories to loan agreements, and as guarantors from whom the lenders may seek redress in the event of a default in repayment. At times of need they have even pledged humans themselves, specifically promising daughters in marriage to others who lent them food, money, or animals to get by. But there is one thing Luo people have long, or at least until very recently, refused to put up as security for loans, or to allow to be so offered. This is land, and in particular the land that one has received as patrimony or in which one otherwise is expected to share claims with other people. Studies and interview reports from the early and mid-twentieth century (for instance, those reported in the Kenya Land Commission in the early 1930s) are clear and consistent on this.[32]

As foreign bankers and foreign-influenced government lenders have introduced loan schemes into Luo country, they have tried many kinds of loan security, including crops, machinery, buildings, salaries, and land; and they have tried various sorts of third-party guarantors, as individuals and in groups. In the cases where they have attempted to base lending on land title collateral, countless misunderstandings and hardships have resulted in the Luo country and elsewhere in western Kenya. In large part, the hardships arise because the use of land to secure loans of cash, farm supplies, or other things neglects the attachment that Luo people feel to the particular lands around the places where their ancestors lived, farmed, died, and are buried, and around which their kin and quasi-kin reside. Many Luo and other East African people would rather fight than give up these lands, even when they have agreed (by their own volition or others' inducements) to sign or thumbprint them over as collateral against loans, or when other kin have done so without their consent. Luo thus reveal their ties to land, and to kin through it, to be among their most sacred concerns.

To sum up this chapter, several points emerge. Feelings about entrustment and obligation are inextricable from feelings about social proximity and distance, and these are culturally specific. Indigenous rural "moneylenders" as such are uncommon in the Luo country, but some lending in cash goes on anyway, some of it in very subtle forms like school fee payments reciprocated only years later or indirectly. Much borrowing and lending takes forms other than money. Like people everywhere, Luo expect to exchange particular kinds of goods and services on particular terms in particular kinds of relationships, and loans—exchanges reciprocated over time—are just one piece of a much broader picture. Intimacy is possible to measure in some ways—for instance, by type and number of kin ties separating a borrower and lender. This sort of formal reckoning does not always neatly coincide with people's feelings of intimacy, since friendship follows its own obscure rules, and since even kin who are siblings may or may not behave like the kind of "brothers" or "sisters" they may ideally conceive. Nor do the closest kin, even when they like each other or know each other well, necessarily trust each other with the easiest loans—for instance, with money, since potential lenders know it can be socially awkward to call such loans in. But the greatest misgivings of all concern loans between strangers, or near strangers, involving land. The idea of securing loans with collateral is not new to Luo, but nor do Luo deem just anything usable this way. To attach the wrong resource in a loan between strangers is to challenge deeply rooted assumptions about human responsibility.

CHAPTER 6

Marriage on the Installment Plan

The Present and the Promised

Trust follows bridewealth.
—ACHOLI PROVERB

Marriage in equatorial Africa is not just something that happens in an hour. It is a protracted process that can take longer than a lifetime, and one can be more married or less, depending on where one is in the process. A critical part of that process, traditionally, has been the conveyance of marriage dues, or bridewealth, going mainly from the groom to the bride's kin.[1] This not only takes time, typically, but also directly or indirectly involves many people. There is enough back-and-forth in the process so that both a bride's and groom's kin must take turns playing host and visitor, giver and taker.

Of all the different kinds of entrustment in which Luo engage, none has been more important, economically or symbolically, than these dues or "payments" involved in marriage. The debts and obligations these involve could be called a central pillar of Luo and East African social organization.[2] They tie in tightly with indigenous and exogenous models of the segmentary lineage system, and with polygyny (see Parkin 1978). But they can also be counted among the chains of women's subjugation, and the topic of marriage dues is one of debate and of custom sometimes honored in the breach.

The Nilotic proverb "trust follows bridewealth" can mean several things. Any exchange involved in marriage may help cement some sort of social bond, a basis for trust. Such bonds may take simple or more complex forms. The simplest form is the merely reciprocal, but even this can involve time and a kind of indebtedness. A saying Gordon Wilson heard among Luo, to the effect that "the obligations of bride wealth never finish," recalls Goethe's words in *Elective Affinities:* "The sum which two married people owe to one another defies calculation. It is an infinite debt, which can only be discharged through all eternity."[3] But to leave it like this, as though no one but two people were involved, would scarcely describe marriage obligations in a Luo or East African context. Marriage here is about others too.

Everyone born in the Luo country is a potential giver and recipient of marriage dues or bridewealth. But usually one will give and take at different times. What A gives to B, A expects to receive back in rough equivalent from some third party C later on.[4] Hence "A Luo man pays dearly for rights in his wife, including all rights to children born to her. He is then hardly likely to forgo the 'repayment' of bridewealth one generation later at his own daughter's marriage to a fellow Luo" (Parkin 1978, 246). Nor need it even take a generation for a man to play roles of both giver and taker. Young men depend in part on the marriages of their sisters to bring them animals that they can then use in turn for their own marriage. Serial dependencies like these cumulatively help stitch society together, long enduring and serving many purposes.

Bridewealth payments and the ceremonial expenses of marriage are often misunderstood by outsiders—tourists, missionaries, bankers—who deem them simply archaic or irrational. Development financiers in particular, concerned more with their creditors' production than reproduction, and more with repayments than with interfamily relations, voice scorn for borrowers who put their crop loans into cattle (especially for plural marriages), or who refuse to sell their animals to repay their farm loans. They look upon indigenous marriage practice in general as a sinkhole of luxurious spending and consumption. What are these critics missing or misunderstanding?

Marriage Dues, Entrustment, and Counter-Entrustment

Marriage itself is no easy thing to define or compare cross-culturally, anthropologists have found. The term in English can refer to a joining, a blessing, an exchange, or a commitment or covenant. It can mean a performance, a protracted process, or a steady condition. It can imply mating, but mating occurs without it. It can point to a set of rights and duties, including ones that sanction or legitimize cohabitation, sex, or childbearing or offspring themselves—among other things. Marriage may imply domestic union, but married people do not necessarily live together (as so many African migrant workers and their families know), nor can all cohabitants be deemed married. Marriage might imply a mood or affective state, but not one universally agreed. Luo who speak English all know and use the term *marry*. They use it to refer to a man and a woman (or a man and multiple women), but not necessarily: a woman can under certain circumstances be "married" to a ghost, or to another woman in a man's role.

Customs of marriage payment also differ around the world. Bridewealth is typical of Africa south of the Sahara, but dowry (as transfer of goods from a bride's parents to her) more typical of Eurasia. The difference may be due in part to the higher relative value of women's labor for cultivation in Africa, where population densities have been lower but the plow used less (J. Goody and Tambiah 1973). But bridewealth persists in Africa even where population is high, as in western Kenya.

The "customs" and "traditions" described in the following pages have all been documented by careful observers and interviewers, and some observed by me, as regularly recurring patterns in and around the Luo country and western Kenya. In truth, though, it is hard to tell just how consistently some of them have been practiced, or how pervasively, over the Luo country and beyond, since customs and traditions disappear, resurface, and get reworked as economic and political conditions and fashions change. Luo in town or country opt for "traditional" and newer church-based forms of marriage (sometimes called "confirming" marriage), or mix the two; some elope or just cohabitate; and a few feign one sort while carrying out another or none. Rare, though, is one who grows old without mating *or* marrying. This one is likely to be pitied and feared. And anyway, the choice of when, whether, and whom to marry is not deemed just an individual's choice.[5]

To most Luo people, up to the present day, there is no proper marriage (*kend,* or *kendruok* as the condition) without some sort of transfer involving animals.[6] Rural Luo, and many urban ones too, make marriage payments partly in cattle, often supplemented with sheep or goats, and partly in cash. There is no fixed rule determining bridewealth quantities: these are negotiated (and sometimes later renegotiated) according to the wealth and status of both the groom's and bride's family, and to the personal qualities and potential of the groom and especially the bride. Conventional amounts have varied greatly over time and space.

Marriage with bridewealth, or marriage dues, is a case of *entrustment and counter-entrustment,* comparable in a way to a pledge. Neither does a bride's natal family relinquish all claims over her as a marriage gets under way, nor do the groom and his kin relinquish all claims to the animals and other goods they offer in compensation for her. If something goes badly wrong in a marriage, each side will want something back. This dimension of counter-entrustment, formally symmetrical in this limited sense, makes the custom an interesting background against which to consider other forms of exchange involving counter-entrustment, notably the pawn and mortgage, being treated in a related work. In the case of the mortgage, the risks that one side accepts are of a completely different order than those accepted by the other and there is a ticking clock that puts pressure on one side only.

Whether bridewealth-using people in Africa think of marriage transfers as "payments" — that is, quasi-monetary economic ones, is something easy to debate and to misunderstand.[7] The short answer is: not necessarily, but they can. Luo sometimes say that a human being cannot be bought or sold, and that bridewealth is an expression of honor and thanksgiving (*ero kamano*), not a price. The Luo verbs commonly used for handing over bridewealth (or marriage dues) are *nyombo* (v.i.) and *nywomo* (v.t. — it takes the person or the animals and so on as object), which refer only to marriage; and *tero* (v.t.), to lead (the animal[s]), which need not imply sale or connote monetary worth. And yet Luo people of both sexes sometimes speak of marriage in an idiom of exchange that surprises foreigners. People who have handed over bride-wealth or women who particularly resent being "paid for" might use a verb like *ng'iewo*, to buy or sell, but the bride's parents, interested in both con-tinued bridewealth inflow and residual claims over their daughter, are un-likely to acquiesce to such a term, as Emin Ochieng' Opere pointed out to me in 1991. Marriages indisputably involve economic debts as well as sym-bolic ones. Goods and services passing from the groom and his natal kin to the bride's natal kin are conceived of, at least in part, as compensating for a woman's labor, sexual favors, and not least important, childbearing in the past, present, or future.

This last point is important. Luo *bridewealth* and *childwealth* — that is, the transfers of wealth for *uxorial* and *genetricial* rights, access, and control — are usually understood in Luoland to be one and the same at the time of pay-ment. The same set of terms — *dhok keny* and *dho i keny* (marriage cattle), *pesa keny* (marriage money), and the generic term *nywombo* — are all used to refer to both.[8] Only in the case of a divorce after childbirth, it seems, do bride-wealth and childwealth get sorted out. In that case the bride's natal kin are expected to return the bridewealth with a deduction for childwealth, a point to which we will return.

Livestock remain important components of Luo bridewealth, even for urbanites.[9] I have known Luo needing to pay bridewealth debts far away to sell their animals, travel with the cash to their brides' home area, and buy other animals there to present to the bride's father and kin.

Bridewealth/childwealth payments are not just payments between individuals. They commonly involve the groom's and his family's begging or borrowing, or both, animals from agnatic and perhaps other kin too. If one of the groom's sisters later marries, the incoming bridewealth may well be claimed in part by those who contributed for him. The current bride's family too, on its side, may be expected in turn to distribute the animals to several members — again, agnatic and other — with claims on them, which may derive from their already having pitched in for a marriage of one of the bride's brothers. That is, both the contributions from groom's kin and the distributions to bride's kin may be parts of long-term entrustments or returns within their respective sides.

Animals and money serve somewhat different functions in contemporary Luo marriage, and they carry different symbolic values. These are expressed in ritual. When Hezekiah Oluoch, a respected elder, died in Kanyamkago in 1991, his son's new wife was not expected or allowed to wail at his funeral (though she could quietly attend) because no cow had yet been paid to cement the marriage; and it was said that no amount of money could have fixed that. Nor are wife-taking affines expected to come to a funeral — they are not yet recognized as *oche*, in-laws — until at least one animal is transferred. But the rules and expectations about cattle and cash are alike enough to keep Luo debating whether cash is substitutable for cattle.

Luo also debate whether any money is substitutable for any other money. On this the consensus of Luo elders, as noted earlier, is no. Only money obtained through morally acceptable acts may be used in bridewealth. Other money — for instance, from selling patrimonial land as an individual, or from selling tobacco or cannabis (crops associated with the ancestors), or from bribery or lottery winnings — is *pesa makech*, "bitter money," and must not be used for bridewealth. If it is, one risks losing one's bride, one's herd, or one's offspring and even one's entire lineage as a result, unless these are put through an expensive and possibly dangerous purification ceremony.

Quantities of animals and cash handed over in bridewealth depend on many things, including the location and its wealth in livestock and land, and the desirability of the woman (virginity, looks, skills, education, demeanor, and other factors coming into play), and the capacities of the groom. It is hard then to fix a "going rate" over space or time — it depends. Cash precedes any cows; it is a kind of "negotiation fee" or "acceptance money."

The numbers do not hold steady historically. Early European observers writing retrospectively from oral reports about the period before the rinderpest and pleuropneumonia epizootics of the 1890s describe twenty to forty head of cattle as having been customary. Evans-Pritchard, writing probably on Luo around Alego when he looked back upon it in about 1950, said it had varied between about fifteen and twenty in 1936 but had since fallen to about two to seven (Evans-Pritchard 1965b, 238–39). Potash's survey of 347 marriages in a rural South Nyanza Luo community close to the Winam Gulf in 1973–75 determined a conventional number of cattle as "approximately ten head" (Potash 1978, 392).[10] By the time I arrived in South Nyanza in the 1980s and '90s, a man in Kagogo in Kanyamkago would expect to hand over six cows or fewer, after opening with a few hundred shillings, for a bride with only primary schooling—unless he was salaried, in which case the demands might rise.[11]

If bridewealth amounts have slid, the decline probably results from rising population pressure on land and the falling value of female farm labor, and perhaps in part too from rising school and medical fees, and use of cattle in mortuary celebrations. Adjusting marriage payments in East Africa is an important East African way of coping with food shortages, particularly in dealings between agrarians and pastoralists.[12]

In an economic regard, as in many other societies nearby and across Africa, marriage is conducted on something like an installment plan.[13] Many Luo make their main marriage transfers only gradually over months or years: and a man and woman are not deemed fully married until the last cow or shilling, or combination of them, is handed over. Similarly delayed marriage payments are standard practice in many societies in tropical Africa, though there is much variance in the norms from one society to the next.

Case material and statistics on bridewealth delays appear in Potash's study of Luo, researched between 1973 and 1975.[14] These figures lead Potash to a conclusion consistent with my own impressions: "payments are usually made over an extended period of time and are seldom completed in less than ten years" (Potash 1978, 392; cf. Håkansson 1990).

A central implication of this is a status hierarchy between wife-givers and wife-takers. Since women and their reproductive potential are at least tacitly acknowledged to be more valuable than any number of cattle, and since they usually move to their husbands' homes before all the cattle are paid, wife-givers have the moral edge over wife-takers.[15] Marriage with bridewealth brings to mind the biblical adage that "the borrower is servant to the lender" (Proverbs 22:7), but in Luo marriage this is more an ideal than a real condition, since the norm of clan exogamy means that typically the groom's and bride's families live at a distance.

Luo marriages have not in the past occurred merely in random directions but seem instead to have followed some discernible patterns in the aggregate. In an unusual attempt to map out the interactions on paper, by A. B. C. Ocholla-Ayayo (1979), Luo in southern and central Nyanza appear as five "blocs" of four or more *ogendni* (what I have called clan federations) apiece. These vary in the degrees to which ogendni in a block intermarry among themselves or instead marry other blocs. And between any two blocs, women tend to move in one direction and cattle in the other. On a larger scale, oral histories suggest that Luo have tended to take in more wives in marriage than they have let go to other ethnic groups, which suggests they may have given out more cattle than taken in, but successful raiding—once a precondition for a young man's marriage—may have made up cattle deficits.[16] There are many unknowns in all this, however, and it is easy to overformalize blocs that may themselves have been unstable in their alliances and flows that may have shifted in both direction and strength over time.

Interlinking Marriages

How much a groom can afford for bridewealth depends on who else in the family has recently married or is in line to do so. A young man with several brothers but poor in sisters is likely to have a harder time procuring marriage cattle than one with several sisters and no elder brothers, and his father and other kin are expected to budget their animals carefully, lest the son turn into a cattle thief or turn against his father.

Cattle poverty can force young girls or women into marriage, as the following case demonstrates. In 1970, Simeon Baraza in Kanyamkago wished to marry, and he being in his early thirties—a respectable marriage age for a Luo man—his father Owere agreed to make it happen. But Owere had no cattle to pay the bridewealth, so the latter determined that Simeon's sister, age eleven, would have to be married to provide cows for Simeon. Owere, the father, arranged with his friend and churchmate, Onyango Pala, to marry her to the latter's son. Marriage followed friendship, and cattle followed marriage. Even though the girl was so much younger than her brother, her compulsory marriage was not considered an unhappy ending to the story. What really mattered—as usual in Luoland—was that everyone be able to marry somehow.

Sometimes entrustments and obligations interconnect the marriages of half-siblings, and even the marriages of parents and children. The following case, more complicated than the one above, shows how this can happen.

Meshack Ogutu, a farmer who had migrated from central to southern Nyanza, had after arriving borrowed three cows from a wealthy neighbor, Odidi Okech, to marry his second wife, Pilista. When Meshack Ogutu's first daughter by his first wife married, he received only one cow and a thousand shillings in the bridewealth (why not more cows is unclear to me). For his second daughter he received five cows and, again, some cash. When Meshack's third daughter Adhiambo (daughter of his second wife) reached adolescence, Odidi Okech, his earlier benefactor, expressed interest in his own son Jason's marrying her (which can be a common way of calling in an earlier debt, as noted earlier). Meshack Ogutu and Odidi Okech talked long and agreed. Jason and Adhiambo also talked long and agreed. Odidi Okech and Meshack Ogutu agreed that three cows had been "prepaid" by Odidi's earlier help in Meshack's own second marriage. He paid four more live cows toward the marriage. In 1991, the son Jason was also expected to contribute some himself, but despite his having a steady job as a tradesman in Nairobi, he had not done so yet. Meshack asked his own son, Odede, when the latter visited Nairobi, to forward Jason a request for the next installment of bridewealth cows.

Meanwhile, from his earlier inflow of cattle from first daughters three, Meshack Ogutu was able to provide most of the cattle for Robert, his first son by his first wife. Meshack's brother Calestous also provided one cow (Meshack's first wife and Calestous's were sisters, and the two families were very close.) Some time passed, but Robert's wife conceived no child—a personal and familial sorrow in Luoland. Robert left her in his paternal home— evidently just to find a job—and went to live in a nearby town, where he was employed as a civil servant. But by the time Robert expressed a wish to marry a second wife, Meshack's second son, Philip, was close to marriageable age. Meshack, having few cows and faced with a tough decision, was either unable or unwilling to contribute any more cattle for Robert's second marriage, deeming this a luxury rather than a necessity at this point, and perhaps supposing that Robert's government salary (and other earnings?) would cover Robert's second marriage. In this he proved right; Robert managed a second marriage on his own. Meshack provided some animals for Philip and Onyango, the second and third sons, for their marriages in turn. For the marriage of the latter, he used some of the cows from Adhiambo's marriage, even though Adhiambo and Onyango were children of his two different marriages. Then it came time for his fourth son (yet another son of his first wife),

Odede, to marry. Having already expended nearly all his cows, having paid for all of Odede's school fees, and having three sons left behind Odede still to marry, he would leave himself with no cows left if he paid what it would take now for Odede's marriage. So for the time being, he did not. Instead he allowed Odede to bring his intended bride into the homestead for a trial marriage without making a bridewealth payment. Even though Odede stood to gain cows for his own marriage by forwarding his father's request or demand to Jason, Odede confided in me that would be the last thing he would ever do—he had too much self-respect, and anyway his bride, a highly schooled young woman, had her own reservations about the whole tradition of bridewealth and its implications for a woman's position and dignity. At the time when I left Kanyamkago, it was uncertain whether any bridewealth would be paid for Odede's marriage at all. He looked likely to need help from his elder brothers, a lucrative job or a loan—or maybe all three—to be able to swing it. Already the elder (half-) brothers were concerned about what Dunstan, the infant son of Meshack Ogutu's second wife, would expect for his marriage when he grew up. They suspected he might feel entitled to just as many cows as his elder full sister, Adhiambo, had received for her marriage— four, plus perhaps also the three "prepaid" for her by the original loan from Odidi Okech to Meshack Ogutu. This seemed likely to pose a hardship for the half-brothers, the sons of the first wife, in the event they had not by then completed their bridewealth. And in all likelihood they would not.

The case shows several features of Luo kinship and marriage. It shows the need for bridewealth, and the ostensible repayment of humans with animals *and* of animals with humans. It illustrates the need for sisters to provide marriage cattle for brothers, and the potential for rivalry between the offspring of co-wives. And it shows the stretching of bridewealth debts over generations—here with much mutual forbearance and forgiveness.

We look now at some of the transfers, rituals, and visitations involved, making some effort to sort out older from newer variants but paying attention to how many of the basic expectations seem to change little over time.

Like an Extended Dance

Cattle and cash both enter into a complex series of transfers and counter-transfers that only gradually cement a marriage. The process of marriage is not a simple exchange of women in one direction, and animals and money in another. It involves a cautious and rhythmic enactment of giftlike transfers back and forth between two networks of kin, and between members of each of these internally. It is a kind of dance—in a way almost like a square dance—extended over time and space, a dance whose nature is dramatic as well as social and economic. The unwritten choreography involves a kind of sacred sequence. Luo call all this *meko*, proper marriage, or *keny mogwedhi*, the blessed form of marriage.[17] What each party owes another always depends on where they fall within a longer course of events and, along another dimension, within longer strings of attachments.

Present-day Luo, when speaking loosely of the past, say that marriages were all arranged by parents and by other kin acting as go-betweens (*jogam*, sing. *jagam*). In classic texts, both bride and groom had their go-betweens, both eventually compensated in meat from animals ritually slaughtered as part of marriage (see, for example, Mboya 2001). In the mid-1970s, Potash's survey of rural Luo by the southern side of the Winam Gulf determined that 273 (79 percent) of 347 marriages were "arranged," whereas only fifty-two (15 percent) of the couples "met on their own" (Potash 1978, 381). The proportion of couples who meet on their own may have risen somewhat, but the line was never clear-cut; many cases involve some prior acquaintance before a parental agreement and thus fall somewhere between "arranged" and "free" marriage. Families and *jogam*, go-betweens, have always been somehow involved, but it has long been possible for some individuals who wanted to thwart their marriages to do so.[18]

In any case, young men and women quite often make acquaintance now in marketplaces, schools, plantations, or other settings outside home and beyond family control. They may and some do date and have sexual relations on their own, and even keep house jointly, without being deemed married. In practice, then, the line between unmarried and married itself is far from clear. Couples can live together in trial marriage, married ones apart in town and country homes. But where marriage and offspring are in question, so by tradition is the exchange of wealth, and this means entrustments and obligations.

To follow a step-by-step enactment of the protracted series of rituals in "traditional" Luo marriage could take a volume in itself, and there are several published ethnographic accounts available that all make shorter attempt.[19] Here, for brevity, only a few points may be made. The ceremonies, involving many animals, can be racy, splashy, and sometimes to contemporary eyes rather sexist.[20] Many of the events are known by special terms (*meko, diero, lupo, duoko, riso*).

The negotiations about whether the marriage will occur, and on what terms, can involve dozens of people on both sides (Odaga 1971, Cohen and Atieno Odhiambo 1989), and they can involve a great deal of posturing. The visitations back and forth between bride's and groom's families again involve many people. They also involve slaughter of cattle, sheep, goats, and roosters or chickens—not just one but by both sides, and the sex as well as species of the animal often specified. Adding show-off foods (and here wheat bread and butter may count as exotic delicacies) adds to expenses of the entertainments.

The initial enactments involving the bride and groom themselves can also involve plenty of theatrics, including an abduction by a groom's delegates, even a mock skirmish during which a bride is expected to struggle, if only to preserve her family's dignity by proving she cannot be taken away easily (and in some cases, her reluctance is genuine). Symbolic apparatus like cattle ropes and a *chieno*, a tail-like tassel, are used to make statements about a bride's domestic duty and fidelity—without always "tying" or animalizing a man in like fashion. A groom must pay for the privilege of sexual intercourse (with a *dher gonyo chip*, the "cow to remove the pubic cover"). A witnessed ritual defloration tests a woman's virginity; and if she fails the test, the groom will owe less bridewealth for her. A ceremony called *riso* with ritual commensalism, involving a bride's and groom's sharing and feeding each other a handful of goat or other special meat (as done elsewhere with wedding cake), is a binding step to a marriage, enacted before or after the completion of bridewealth transfers (so one can be fully married in one way but in another way not). Riso affords an occasion for unusual sexual license, in which unmarried kith and kin of bride are invited to visit and cavort at night (though stopping just short of full sexual intercourse) with those of the groom. There may also be a longer period in which a new bride (*miaha*) may be given extra sexual liberties around her new home.

But the main burden of the bridewealth, as far as sexual relations are concerned, is to tie down the woman more than the man. Sexual relations outside the marriage are conceived of as more of an infidelity and a breach of custom if carried out by a wife than a husband; indeed this sort of behavior has been practically expected of men—more or less consistently with the male ideal of polygyny—at least up to the era of alarm over AIDS and other sexually transmitted diseases in the late twentieth and early twenty-first centuries. In private everyone knows that very young brides married to very old men—a concomitant of polygyny in itself—are likely to have sexual relations with other men than their husbands for both recreation and procreation. The bridewealth means that they are still contravening something when they do, and most people involved seem to live with this cultural ambivalence.

These, then, are some of the elements and activities described as conventional—or at least conventionally described—as making up "traditional" marriage. While often spoken and written of as a sequential package—and indeed the order of enactments is crucially important—in practice it may be better understood as a repertoire with only a few essential points (for instance, the meko with initial bridewealth payment and the later riso).

But let us return to the debts and obligations involved. If a groom's father and his kin approve of the woman, he offers his son one or two head of cattle —sometimes more—to be taken to her mother as a starter bridewealth payment. This is an obligation in itself, and it connects one to generations before and after. As Emin Ochieng' Opere explained to me in 1991: "Because if the old man fails to do that [provide the initial bridewealth cattle for the son], then it is a debt on the old man's side. You see, you will later on, when you grow old, you'll say, 'My father never assisted me in getting married.' It would be very bad, and it may even result in that man or the lady being married, failing to assist other people in future."[21]

Luo nowadays conventionally pay the initial bridewealth in both animals (including a cow) and cash. The cattle are most often young males or females. The cash is a token of agreement called *ayie*, in southern Nyanza (perhaps a cognate of *ayieth*, meaning a rope loop used to tie a cow to a stake). In 1990, *ayie* in southern Nyanza usually ranged between Ksh. 1,000 and 10,000 (or U.S. $42 and $420), depending on both the capacity of the groom and the qualities of the bride. Of particular importance in this regard is the bride's education, seen as a store of wealth, among other things. As Emin Ochieng' commented in a 1990 interview about an unfolding marriage between a university-educated man of Kanyamkago and a young woman who had completed secondary school, "You see, Jeshon can't take [just] one thousand to their home, because the parents, I mean, sacrificed to educate her up to that level of education, Form Six . . . remember the bride's mother actually toiled. So maybe he will pay something from 3,000 onward . . . she's consumed a lot of her parents' wealth."

So bridewealth, a "traditional" exchange, ties directly into school fee payments, a "modern" one, in a clear understanding of humans as stores of monetary value and of potential earning power: what some would call "human capital." If a bride's parents and other elder kin, knowing she will marry away, labor or sacrifice for her schooling anyway, it is not just with the altruistic intent of improving her own life chances but also with an expectation of a return for themselves: *an entrustment in a process.*

Earmarking, and Bridewealth's Branching Pathways

We have seen instance of "earmarking" in the choice of certain money and animals, acquired by moral means, for use in bridewealth. Now we see another. Bridewealth cattle include animals given obligatorily to the bride's father for passing on to other kin (these animals are called *dho keny* or *dho miluhini*) and others paid to him alone (*dho i keny*).[22] Animals of the former sort, given first, are designated to go to particular relatives once they reach the bride's home.[23] Evans-Pritchard's description from Alego in 1936 (1965b, 239), of a two-year initial sequence of *dho keny*, differs only in details from that printed in Gordon Wilson's manual of Luo marriage laws based on a wider area. The latter specifically mentions this sequence of animals destined for relatives (*dho miluhini*, or *dho choke*) (Wilson 1968, 114, para. 72):

- *Dher maro dwasi gi nyathine:* a milk cow and its calf given to the bride's mother;[24]
- *Dher wuoro dwasi gi nyathine:* a milk cow and its calf, or a heifer and a bull, given to the bride's father;
- *Dher owin wuon nyako* (or *dher atung'*): a heifer for the bride's father's brother;[25]
- *Dher omin nyako* (or *dher chiewo*): a milk cow or heifer and a bullock for the bride's half brother; and finally, after the groom has been feasted,
- *Dher ner nyako:* a heifer or a bull for the bride's mother's brother, the last in the series *dho miluhini*.

As the lineup indicates, animals of specific ages and sexes are destined for particular kin—just as, upon ritual slaughter, specific *parts* of a killed animal are also designated for particular kin.[26] Some of these animals may, of course, be subject to further claims upon those recipients by other kin in turn. The sequence above is (or was) kept partly so that if the marriage fails, the animals can more easily be accounted for and returned. It is all conditional.

The second phase of bridewealth, *dho i keny*, involving animals for the bride's father but not earmarked for other kin, may take many years longer—maybe until the wife's daughter marries and it becomes an issue of rushing to complete the earlier marriage to keep things in proper sequence. How hard a wife's father presses his claims is likely to depend on his personality, relative need, and assessment of the marriage.

Nowadays, as grazing land has become scarcer, titles are issued, fences are going up, and herds are becoming harder to maintain, customary bride-wealth amounts have shrunk in Luoland. So not all these kin of a bride may get what would once have been their share. Still, the basic principle applies: human ties and animals partake of each other as always. Even today, one who wishes to market an animal for slaughter, or to repay an institutional credi-tor such as a bank, may well be seen as betraying a kinsperson with claims to the animal.

What Gordon Wilson wrote in the mid-1950s, that "it usually takes the average Luo many years to pay off all of these animals" (of bridewealth), remains true today, though now there is more likely cash in the mix (Wilson 1968, 115, para. 76).[27] Rare is the young Luo husband who has fulfilled his bridewealth payments to his wife's father (*jaduong'*—big one, senior one) immediately upon marriage. The debt is a keystone of the authority of bride's father over daughter's husband, as well as of the father's authority over the son the groom (who may be depending on him for more cattle to pay the bride's father). One who feels his daughter's groom is not deliver-ing the goods on schedule can call the bride home; and if she feels loyal to her natal kin or has met problems among her affines, she is likely to go.

Bridewealth payments can involve a certain amount of subtle threats and brinksmanship. A bride's father retains the right to call her home, or at least to invite her often to visit back, if he is not content with bridewealth pay-ments coming in from the groom.[28] (Or at least it is made to *look* like his right and decision: one may suppose his spouse, mother, or other women-folk counsel him behind closed doors.) A young man in my acquaintance had more than once lost his wife to her father's home, as he was having trouble with the payments. When the young man was offered a paying job and the wife's father heard tell of it, the latter's demands for cows only increased.

Letting bridewealth come slow can subtly benefit even the wife-givers, as it keeps their lineage's future herds spread out in more than one locality, cut-ting down risks of disease and theft. Allowing wife-takers to pay up slowly is also a convenient way of keeping one's wealth (or potential wealth) invisible, and thus of avoiding jealousies.[29] A bride may follow a comparable strategy of her own. Typically, early in a trial marriage, she receives some fine cloth-ing from her husband, backed by his kin, who may need to impress her own natal kin in order to keep her. By keeping items divided between the two homes, and lending some to her sisters back home, she can both avoid incur-ring jealousies and cut the risks of losing everything in a theft.

Which cattle to choose? According to Dan Odhiambo Opon, an articulate, university-educated young man whose bride was living with him in 1991 and who was just confronting the issue of when and how to pay his bridewealth, old parents usually advise their sons to choose younger animals (whether cows or bulls) for bridewealth. They do so for three reasons. In his words: "(1) Younger ones are a symbol of beginning of life of a young couple. (2) Any man who takes any adult and fat bulls and cows for bridewealth will be suspected of being either a wizard, a malicious witch doctor or a quarrelsome husband, and the attractive animals are [perceived as] intended to prevent any possible divorce. (3) Neighbors of the bride's parents may feel jealous and play harmful magic on the bride and she may even die before long, if attractive animals are taken as bridewealth."[30]

These words remind us of one of the most important lessons about entrustment: that *there is usually some third party,* if only a silent or imagined spectator denied, whose expected response can condition the terms of exchange between the principals. The presumption of neighborly envy, and of magical sanctions, may be doubted, but the doubt and anxiety probably help keep these perceptions alive.

Bridewealth animals that die the groom must replace, as if by warranty. For proof of death, the bride's father is expected to send him its skin, plus some meat or money from meat sold.[31] Why the son-in-law should care whether the animal died or got sold could only be explained by the likelihood of its dying in the first place; and the need to replace a dead animal would seem to help ensure that grooms not deliver animals they know to be sick. On the other side, traditionally, should the bride herself die young, she too has been deemed replaceable by a sister from her home.

In all the long series of visitations back and forth that can accompany a marriage, families of brides are likely to keep records of their expenses, because in the event of a divorce, these are deductible from the wealth they must return. This is as true today as it was in the 1930s as described by Evans-Pritchard. With or without written accounts, as Emin Ochieng' Opere told me, the woman's family is expected to exaggerate the amounts of food consumed and its cost. Grooms anticipate all this, and whether for this reason or just to avoid all the fuss and inconvenience, they minimize their visits to their wives' parents homes early in the marriage—for instance, by paying their bridewealth animals several at a time if they can. But there have been cases where, after deductions, there was nothing left to be returned.[32]

Divorce and "Deductibles":
Redeeming Marriage Dues and Favors

When marriages turn sour and end in divorce (*weruok*), as suggested above, bridewealth is returnable. It is indeed this return (and all it signifies) that in recorded "customary" law distinguishes divorce from mere separation.[33] The latter is common in Luo country, but divorce—the formal annulment of the marriage—has hitherto been much less common in Luo country than in many other parts of the world, notably North America (Potash 1978; Buzzard 1982, 247 and passim). Having children tends to anchor wives to their husbands and husbands' homes, and into their marriages, since in Luo society, so strongly patrilineal in so many respects, a child of divorced parents has traditionally been expected to remain with his social father (that is, *pater*, the one who paid the bridewealth, as distinct from *genitor*, the biological father), and this expectation has been elevated to the status of "customary law" (Wilson 1968, 141–42, paras. 200–204).[34] Land rights too, vesting formally in male hands, keep women married to maintain access to land. Finally, there are ritual pulls, since ceremonies involving a mother's children's marriages and funerals require her presence around her conjugal home (Wilson 1968, Potash 1978). Most of these things remain true in contemporary urban settings—socially tightly linked to rural ones—in recent times (see Buzzard 1982, 260–65 and passim, on Kisumu).

But divorce does occur, and either spouse may initiate it.[35] Before a quarrel or disagreement reaches this point, the siblings, other kin, and any *jogam* (go-betweens) of the couple, and perhaps clanspersons from both sides, are expected to help resolve it. Their decisions about marital wrongdoing may be considered final; but a case may also proceed, simultaneously or later, in a court of law. In the case of marital wrongdoing like wife-beating deemed cruel or excessive, the adjudicating kin sometimes require the husband to make further bridewealth payments to her kin as indemnity. The implication is interesting. Even after marriage, she remains, to an extent, an appendage and representative of her natal family; and wrongdoing to her calls for restitution to them.[36] She is treated not as a permanent, unconditional member of the husband's family; she is rather an entrustment to them.

Another eventuality leads to a similar conclusion. If a young wife dies, leaving children in her conjugal home, her natal family is supposed to send another woman to care for the children. They must not just because they care about the children but also because they carry an obligation to their erstwhile affines. Particularly if the marriage cattle are not returned—a distinct possibility, as they may already have been disposed of—her family now carries a debt. Marriage is not just a couple's arrangement; it is a continuing series of entrustments and obligations between the broader groups they represent.

When a wife decides to leave her husband, he and his family are likely to try to recover the bridewealth cattle (*wero*, take back, or *waro*, redeem, *dhok manenywomo*, the bridewealth cattle paid), often taking their case to a subchief's *baraza*. It cannot be said that they usually succeed. Potash's survey of a rural Luo community south of the Winam Gulf in the mid-1970s found that of 347 marriages, about 33 cases of separation had yet involved no return of bridewealth, as against only 17 that had involved partial or total return (Potash 1978, 386–87).[37] By definition, most divorce cases, with or without the involvement of government courts, do end in returned bridewealth. But the animals returned are not usually the same as the ones given. The wife's family may well try to give back skinny, bony cows if they can. There is, of course, some risk of shame in this, but in formal reckoning—in Luo custom if not also in the courts—a cow is a cow, and as long as the species and sex are right, the husband's family has little recourse. That is, if a husband and wife divorce, quality and quantity divorce too.

How many cattle the husband's family may collect for a divorcing wife depends on how many children she has borne, and what kind. The differences are telling. If she's had a boy, the woman's family will deduct four head of cattle from the returned bridewealth. If she's had a girl, three head stay. Or so, at least, it was expected in Kanyamkago in 1991. Other, earlier observers have noted different numbers elsewhere, but always with a disparity between male and female.[38] Two conclusions emerge. One is that males are deemed worth more than females, or culturally valorized more, in an economic or symbolic sense or both. (The same symbolic message appears in Luo funeral mourning, traditionally four days for a dead man and three for a woman.) The other is that it was not just the woman as a person who was being entrusted to the groom and his family in the first place but also her reproductive potential. Bridewealth is childwealth too, at least up front; and only eventual divorce calls for a distinction between them.

In my information from Kanyamkago, a man's family that demands back a cow may also demand back its calf that has been born since it was paid over, and some Luo assert *its* calf should be repaid too.[39] Certainly it is best for a suckling calf not to be separated from its mother. Arguably, too, the bridewealth represented not just the transfer of an animal but its reproductive potential too. This concept mirrors the image of the wife herself, since a woman who proves infertile may well be divorced by her husband and bride-wealth demanded back. Life for life, fertility for fertility.

Whether a bride's family bears a debt for animals given in bridewealth depends in part on whether it *can* return them. And this depends partly on class and age. Not infrequently, as Potash notes, a girl or young woman from a poor home is married to an elderly (read wealthy) man, and the animals handed over fairly quickly and used immediately for sale for subsistence or schooling, for her brother's or brothers' bridewealth, or other urgent needs. For such a woman, cattle poverty in her natal home is likely to give her trouble leaving her husband. On the other hand, Potash writes, for a marriage of spouses roughly equal in wealth and status, payments usually start small, and so separation or divorce may not bring about a big debt (Potash 1978, 392). Rich or poor, the recipients may simply have lost or handed away the animals when the time for divorce comes. A recipient who can prove having lost the animals to theft may be excused having to return them, according to Wilson.[40]

So divorce does not necessarily mean a bonanza of returned cattle for a husband.[41] Since childbirth tilts the balance of obligations between families, lessening the debts of wife-givers in the case of divorce, it is common for husbands emotionally estranged from their wives to try to keep home with them anyway. Their sexual coupling in such cases, according to my infor-mants, gets limited to planting times and other ritual occasions like a son's founding a new homestead. (Their not engaging in sex then would risk *chira*, the mystically caused wasting affliction that can come from events like sex out of proper sequence of seniority.) In all, though, the balance of marital obligations would seem to favor keeping families intact.[42]

It would be wrong to conclude this discussion of the entrustments and obligations in marriage without noting how politically charged these issues are among urbane contemporary Luo. I have known young, city-schooled Luo women who loathed the idea of marriage payments altogether. One young bride in Kagan voiced in 1990 the sentiment that the imminent pros-pect of her husband's paying cattle and cash for her made her feel like a mere chattel sold in a shop, like a kilo of sugar or packet of cigarettes.

Class and education, as well as gender, enters the debates. Dan Odhiambo Opon, a recent university graduate in Kanyamkago in the early 1990s, offered the opinion that "today, only the unlearned still take back their daughter who had been married in the event of the husband's failure to pay the bridewealth in time. Most educated couples will decide how and when they should take the bridewealth. A man may be of less income and his wife may assist him to pay the bridewealth. This is a recent development with the deeper penetration of western culture into the minds of the educated lot. The elite in this issue find that marriage should more be an affair between the concern[ed] couples than between their families."[43] What you do about bridewealth thus helps determine, in some Luo minds, how modern, educated, or sophisticated you are. The point here is not so much a new individualism or emphasis on autonomous small families but the *perception* of it, and its presumed association with foreign influence.

Some lament, too, that the old marriage customs no longer hold sway with the same force as they once did; some point to elopement (*por*) and marriage without bridewealth as a sign of social decay. But these things are nothing new. They were deemed a trend even in Gordon Wilson's time just before Independence, and Luo elopement was observed by Evans-Pritchard (1965, 229) in 1936, almost two decades before that.[44]

And so around and about. Historical research on neighboring Gusii similarly shows *episodes* of common elopement, and also bride abduction sometimes including rape, during periods of high bridewealth expectations temporarily risen too high for young men to meet. Daughters whose fathers sought to arrange them marriages to wealthy men sometimes lost them to the men the daughters picked for themselves. This much, the stuff of romantic legends, probably goes on most anytime, anywhere. But abduction with rape seems to have risen in incidence in the 1890s (rinderpest era), early 1940s (wartime years), 1950s, and 1960s and might still continue on occasion. Rather than rejecting marriage altogether, young people appear in these periods to have rebelled in these ways against senior men's control, and probably pulled down their bridewealth expectations. Abducting and penetrating a young woman sometimes seems, by diminishing her value to other men and their kin, ironically to have made her easier for her captor and his to hold onto later. Bride abductors were, and maybe still are, expected eventually to hand over bridewealth to make good (Shadle 2003; see also Shadle 2000, Håkansson 1988). These are some of the ways obligation can arise without an initial entrustment (rather as in the Nilotic story of the lost spear and bead); and in recorded times and places it looks to have happened in waves.

Social and moral breakdown are hard to measure, and it is perhaps a natural tendency to remember past custom or tradition as more consistent than it really was. This may help elders, at least, to purify their own youthful motives in hindsight—as though lust, youthful rebelliousness, or the desire to grow up fast were new in human nature.

Human Pledging

We have seen, in marriage, a form of entrustment whereby the transfer of a woman is followed and legitimized by the transfer of animal and other wealth mainly in an opposite direction. But children, as well as adults, sometimes become involved as principals. A man sometimes uses a young daughter either as a pledge or as an actual loan repayment, if the union is deemed otherwise suitable or just expedient. In 1949, without specifying a period, a Luo man named E. P. Oranga, working for the colonial government, related this aspect of social history.

> In some old days especially during the time of famine, people used to get [a] loan in the form of food crops and in some cases they were using their daughters as means of security. It was a very common practice. Even when one wanted some cattle and he had nothing to pay he would take the daughter to the man who is to give him a Loan and to inform him that if he fails to pay the number of cattle he took, then that man [the lender] would take his daughter and in some cases the daughter stays with him until she is full grown and if he [the borrower] fails to repay, then that man [the lender] adds a few cattle and then marries the daughter. In some cases the daughter stays with the father until she is married and as soon as one begins to pay dowry [that is, a third party pays bridewealth to the borrowing father] then the father starts to repay back the loan.[45]

A daughter is deemed, among other things, a store of value directly relevant to diverse credits and debts.

Sometimes it is only contingency—destitution or other adversity—that makes clear a girl's deemed value. From one male perspective, "The economic value of a girl to society is gauged by the extent to which a man may be excused his debts on the ground that after his daughter's marriage he will be able to repay them" (Ominde 1952, 36). Wilson's page-long account, written in 1955, just a few years after Oranga's and Ominde's, was headed *Nyar osiep*, referring to "child marriage" or "child betrothal," or more literally to a girl so involved.[46] The economic angle shows up clearly: "In [some] circumstances, a man may have cattle belonging to somebody else and may be in a position where he is in dire need of an animal to repay a debt or to contract a marriage. He may under these circumstances betroth a young girl to the owner of the animals or to his son, in order to receive the first animal of the bride wealth immediately." Or the fathers of a young boy and a young girl contract a marriage involving animals already being held in trust by the girl's father for the boy's, if the former has shown the skill and care to increase the herd and would like to gain permanent claims over the animals. In such a case the fathers might exchange drinking reeds (*kiseke*) at a beer party and the young girl thus promised becomes known as a *nyar kiseke*, daughter of the drinking reeds (Wilson 1968, 121, paras. 101, 102).

Note that, in these forms of marriage, one could contract debts and obligations through the actions of one's parent before one had reached an age of responsibility. Moreover, in the case of child brides (or grooms?), it was expected that the long sequence of rituals and ceremonies of marriage — with their many participants — be observed with extra care to prevent *chira*, the dreaded illness that can threaten a family or lineage.[47] For entrustment and obligation are not just individual matters.

As a Kanyamkago informant told me in 1982, if a man has betrothed a daughter to a creditor's son against a loan (in that hypothetical instance, a loan of a cow) but she instead marries someone else of her own choosing, the father must refund the debt immediately — in narrowly economic terms, the loan "security" would be gone.[48]

Pledges of young women and girls can involve the dead as well as the living. If one dies unmarried, her father may give her body away to another man to bury away from the giver's homestead.[49] The man's accepting the charge is deemed a favor to the father and other kin. If the giver later has another daughter, he is expected to offer her in marriage, against a reduced bridewealth, to the man who took charge of the first girl's body. The bridewealth "discount" would seem a compensation for the man's having taken on an unpleasant task with some spiritual risk attached.

A man forced to pay a bridewealth debt may call in turn upon his own "wife-taking" affines (in-laws) elsewhere to pay another installment to top up the family herd.[50] Debts stretch around the countryside, and into the cities, in chain links. Where persons or families have multiple bridewealth relationships, involving both reciprocal and serial obligations, these would seem to provide some elastic in the life of an economy: some way to stretch resources and capabilities over time, and spread them between homesteads in a way that allows give and take, forgiveness or call-ins, in times when adversity hits one party or the other unpredictably. The webbing of debts and obligations, reinforced by all that is sacred about marriage and its emotively powerful rituals, would also seem to form some basis for society's cohesion.[51]

Not that all strands in the web are harmonious or anxiety free. Co-wives, for instance, can get along warmly and intimately or compete fiercely. Their jealousy, the *nyieko* from which their relationship takes its Luo name, manifests itself at times in material entrustments in both city and country, and these can be sexually exploitative. Shirley Buzzard offers the following vignette from her work in Kisumu. Thomas lives in Nakuru but has two wives in Kisumu, both whom he visits monthly. The wives are jealous rivals who avoid each other. But Thomas borrows money from one or both to spend in Nakuru. Each wife fears that if she doesn't lend to him, he will favor the other wife with his "time, money, and land, which are finite, and the approval of his family which is capricious" (Buzzard 1982, 114). Buzzard's study concludes that economic advancements made by women who found wage work in towns had not, by the time it was undertaken, markedly improved their social position. Male control over their lives, and a shared ideology of sexual hierarchy tied to bridewealth and the view of women as property—something that only intensified as cash entered bridewealth—simply remained too strong, in city as in country. But that need not mean forever.

Nor have patriarchy and patronage ever been something that adept women could not use to their own advantage. Luo, Luhya, and other western Kenyan women, including widows making public expressions of grief in funerals, have long been able to manipulate what men deemed a typically feminine condition of emotionality, suffering, and meekness, and a male self-image of strength, into a strength of their own. The patriarchy and patronage have never gone away or been easy for women, but some of these can at least use big public events as their sounding boards in exploiting it (Mutongi 1999 and Abwunza 1997 on Maragoli; see also Hay 1982 and Potash 1986 on Luo).

Mating without Marriage: Commercial Sex and the Threat of Betrayal

Mating is not the same as marriage. They just intersect, as noted, and as some define them one subsumes the other. Not yet thoroughly studied in eastern and central Africa, but surely coming up as a topic connected with public health research, are the fiduciary dealings in short-term liaisons and longer-term unions (quasi-marriages, if you like) entered for sex, money, or food; for lodging, transport, or affection; or just not to be alone. These are the affairs of passion or mere expediency. They can be private affairs, more private anyway than African marriages involving feasting, serial transfers of animals, and all just discussed. We therefore cast the net a bit wider, beyond the lake basin, drawing in some "fiction" as well as non-, since as much truth on this topic can be found there as anywhere. Doing so may eventually add to our emerging understanding of entrustment and obligation, as well as that of the parameters of life for East African people: some of their highest hopes (procreation or recreation; belonging, wealth, or power) and worst fears (mortal illness, childlessness, and betrayal). I hope the topic may be broached without casting moral judgments for or against the persons described.

More is known on heterosexual than on homosexual unions in Africa, the legal and religious strictures being less severe and the secrets less carefully guarded. Male sex trade, for male or female clients, is very often put down to foreign influence—not the whole truth, surely, but an attribution with at least a kernel of truth in it (Carlebach 1980, 76, on Nairobi). Then again, heterosexual commerce is also tied to foreign influence in more than a few angry African works of art and science, and indeed in popular lore such as the Luo image of the "schoolgirl in a hurry," common since the 1930s and '40s (Cohen and Atieno Odhiambo 1989, 96), or of the tourist tempter downtown.

Settling on terminology is difficult. Slang, colloquialisms, and their translations change often and can easily offend. Persons called "kept" women or "concubines" (that is, live-in companions) often turn out not only to provide services that tacitly earn their keep, if not also money and provisions from other earning activities practiced simultaneously, for instance trading. Many disparaged interactions are episodes or contingencies, not necessarily permanent identities, and hurtful epithets for these or their players are seldom fair descriptions of character or full pictures of complex relations. If there is anything that can be safely said about unmarried sexual unions for compensation, it is that they vary widely. No single term like *prostitute* or *hustler* suffices to cover the types that in Swahili divide into terms for streetwalkers (*watembezi*, from *kutembea*, to walk around or wander), women actively soliciting from their doorways, windows or porches (*wazi-wazi*, lit. "bare-bare" or "open-open"), or women operating unadvertised from their homes in sometimes more enduring relationships (*malaya*, lit. "prostitute") or with partners in live-in concubinage. To which sorts one might add bar and dance hall regulars (relatively common in Kenya), brothel workers and managers (less so), women who work in league with taxi owners, and, especially where public solicitation laws are enforced as in much of Europe and the United States, "call girls," actually usually women (Carlebach 1980, 75–6; Muga 1980; Southall and Gutkind 1980, 53–60; White 1990). These types overlap, though, just as co-workers in a bar or brothel may fall partway between "individual" and "organized." Sex work in Kenya and East Africa has tended not to be controlled by large underground syndicates as it has in some other parts of the world. It is less centralized but no less important in its economic or epidemiological dimensions.

Live-in lovers, pay-as-you-go partners, and short-term commercial sex workers (or prostitutes) rely on varied sorts of reciprocity, whether the material payoffs come in cash or kind, or by the season, month, or hour. Women and girls in such ties usually rely on faith, trust, or confidence to a degree, even when operating by the night or hour. The kinds of *betrayal* feared and experienced are also many, including the medical (not revealing knowledge of infection or illness), informational (letting out a secret, or charging to keep it), and simply economic (refusing to pay up)—to say nothing of the possibility of rape or other violence. Historian Luise White studied "prostitutes" (watembezi, wazi-wazi, and malaya) who had provided men with not just sex but also a wider array of "the comforts of home" in colonial Nairobi from the great 1930s Depression on. Many, first encouraged away from home by want, were getting their rural fathers or husbands back on their feet economically, efficiently becoming independent female heads of household themselves, or both. Some women accepted payment after encounters or compensation at infrequent intervals. Others, including most of the ones of wazi-wazi sort, ones working in straitened times of high rents, or ones who had been let down, required payment in advance, stating, "I had no reason to trust the man" or "I did it so no one would cheat me" (White 1990, 84, 89, 106). Women had some leeway to shift the expected sequencing as the acquaintance and other circumstances seemed to warrant. In cases of alarm for nonpayment, threats, or violence, these women did any of a number of things, including calling on known landlords or police (ironically, since their trade was formally illegal), or raising an alarm to other women sex workers for whom they were ready to do the same. Which they could do depended on which of the types of sex work they were doing. At times, though, after being let down by a client who would not or could not pay, they simply let him go—whether out of disgust, forgiveness, exhaustion, or just calculation that the payoff was too small to risk violence or a scene (White 1990, p. 84 and passim). Not that none could remember or talk later.

Or dream of a little revenge. Lang'o Ugandan novelist Okello Oculi's vivid, empathic portrait of a Kampala prostitute and her thoughts, published shortly after that country's independence when he was twenty-six, laments both foreign influences and people who have sold out to them. The narrator abjures metals, not least money, as killing the people and driving them mad, tempting them away from "where the bones of their grandmothers are buried" to fly abroad to look for it. "White people have lied to us . . . to get money out of our hands and lands. What is in this money, people! For us we go here, open our thighs, go there, open our thighs, like dogs urinating on road corners and stumps to mark guides for their noses on strange roads, and all that for money" (1968, 79). But even the rich and powerful sometimes let down their partners and refuse or postpone the expected payments or gifts. What, then, to do about it? One shames a client by exposure. Oculi's conversant ridicules a government official with driver and limousine, a regular client who, every time he has finished with her, "fumbles all over his pockets and then says he forgot the money pocket in the office." She says, "One day I will make him forget his trousers in my place . . . [like] those two girls who did it to some white man. . . . They scooped out his pockets as he snored away." She continues, relating that when he awoke to find them empty, "he ran to the police station with his thing waving 'good morning' to everybody on the road! Ah, this is life!" (Oculi 1968, 54; see also Mwangi's 1976 novel on Nairobi). Here East African resentments of sex, age, race, and class combine to burst the dam of tolerance. And it is up to the human imagination — of betrayed female narrator and male author clearly no less indignant — to turn the pocket inside out: to salvage the bitter betrayal of "the life" into "life."

Outcomes in commercial sex work differ in many more ways, though, than just getting paid this or that time or not, or than acquiescing or pulling pranks to get even. Karen Flynn's study of Mwanza, Tanzania, across the lake from similar-sized Kisumu, shows just how meagerly paid and exploited sex work in a provincial city can be, barely meeting subsistence for street-dwelling women and for girls, often battered, bullied and by her account miserably needy and insecure, yet not merely passive, at the bottom of a hierarchy even of homeless youth. Flynn's study shows how public moral opprobrium can materially threaten survival as well as self-esteem. Another extreme is vividly illustrated by the rare autobiography of a cosmopolitan, itinerant West African *ashawo* (here, "bar girl" with longer-term relationships too) with the pseudonym Hawa, transcribed by John Chernoff. Hawa shows how one can maintain a high standard of living, adventure, and entertainment — at least one of an attractive enough appearance and energetic age. She also demonstrates how one with an unusually buoyant character, a keen interest in the human drama, and a wry, proud humor can help cope with and transcend the betrayals and other shocks still sometimes hard indeed (Flynn 2005, chs. 8–9; Chernoff 2003). Less attention has been given to men's than women's sexual lives, but something similar applies. Variety is the word, both between "cases" and within.

It is not just individuals who are the "cases," and not just women or girls. To the extent generalization is possible at all, sex work seems more often than not to involve differences of wealth and status between male and female partners. Reliable numbers are hard to get, but here are some figures for what they are worth. A Kenyan and international public health survey team visited four rural towns and surroundings in Western Province (Busia, Mumias, Nzoia, and Webuye), mainly Luhya country, in 1999–2000. Its interviewers surveyed 368 "female sex workers" (estimated as about a quarter of the ones working around those sites, and chosen by a networking or "snowball" method). They also surveyed 817 "male company workers." Fifty-seven percent of the female sex workers were based in bars or lodgings, 41 percent at home, and 2 percent in the streets or at a factory (U. Schwartz et al. 1999–2000, p. 12). Fifty-eight percent were Luhya, 17 percent Luo, 10 percent other Kenyan, and 14 percent Ugandan (p. 12). Some of the details are telling, suggesting many of the women were practicing sex among other activities as a survival measure. With a median age of 26.5 and a range from 15 to 50 (p. 12), the "sex workers" in the study reported themselves as 68 percent separated, divorced, or widowed; 31 percent never married; and about 1 percent married or cohabitating. They indicated in 72 percent of cases, without scripted answers suggested, that the main reason for starting was economic or poverty-related. (Another 15 percent cited peer pressure, and 13 percent the draw of sex—maybe the hardest thing to admit.) The picture of marital breakdown and poverty inclining women into sex trade agrees with others' findings since the colonial period.

The takings were modest, by all signs. About 71 percent of the sex workers reported being paid in cash *and* in kind, the median cash portion of the earning reported to be 500 Kenyan shillings (then about U.S. $8.50) per week from sex work, and the most common cash fee per client (per visit, presumably) being between Sh. 100 and 200 (about $1.70 to $3.40). Over 80 percent were found to do other things too for a living, whether bar work (for some, surely a related activity), restaurant serving, trade, or office work (pp. 13–14).

Certainly there are health consequences where people must trust their sex partners not to be carrying infectious diseases. The women surveyed as sex workers were reported to have from one to 6.5 clients (site averages, evidently) with a median of 1.5 per week; and 76 percent also reported having other steady nonpaying sexual partners (boyfriends, husbands) (Schwartz et al. 1999–2000, 17). That many of their partners were acquainted with them seemed to the researchers to inhibit the sex workers' willingness to protect themselves with condoms (or, one may add, to protect those partners). Western Kenya was a region with widespread public information about the dangers of HIV and other sexually transmitted diseases, and most claimed to understand the risk of infection even in a healthy-looking person; but only a minority appeared to suppose their own partners infected. Some 54 percent of the sex workers reported always using condoms with paying clients, but only 22 percent said they always did so with their nonpaying partners. To approach it from the negative, 62 percent reported *never* using them with their nonpaying partners, as compared with the 21 percent who reported never using them with their paying ones (p. 23). Asked why they did not use condoms consistently with a paying partner, the female sex workers most often cited client refusal (Sw., *"mjeta hukataa"*); but asked the same about a nonpaying one, they most often reported trusting him (Sw., *"na mwamini"*) (pp. 19, 22). Lest too much be read into these interesting responses, a look at the questionnaire shows each of these most common two responses to be the first options listed under their respective questions and read out to interview subjects (pp. 43, 45). Taken together, though, these findings do raise a question about the risks of familiarity and trust for further study. Some 24 percent of the sex workers reported themselves to have experienced at least one episode of what appeared to be sexually transmitted disease over the previous year (p. 29).

The 817 "male company workers" in the related survey add to the picture. Twenty-five percent of these reported having more than one sexual partner. Female sex workers were found on the whole younger and less schooled (as well as more mobile) than company workers (Schwartz et al. 2002; see also Luke 2006 on Kisumu), a finding that squares with most reports on commercial sex, and which suggests asymmetric power. Such power difference may help explain why the female sex workers trusted their partners as much as they did in deciding about contraception, though not why the partners trusted them. Here both familiarity *and* power seem to intercede between knowledge and action.

Further information on exchange, sex, and condom use comes from Nancy Luke and collaborators, who studied men and their reports about their sexual partners and encounters in Kisumu, the biggest mainly Luo-speaking city. A sample of 2,700 randomly chosen men, aged twenty-one to forty-five, in 2001 yielded information on them and on the (up to five) most recent sexual partners each claimed to have had in the previous year. The 1,028 men who reported themselves sexually active in the previous month had had 1.5 "nonmarital" but "not commercial" (that is, not with short-term "commercial sex workers" or prostitutes) partnerships on average: these ties lasted on average about fourteen months. Men sampled averaged twenty-six years old, their female partners twenty. Marital status differed too: the men claimed to be 59 percent single, 37 percent currently married, and 5 percent currently divorced, separated, or widowed; the women reported to be 85 percent single, 4 percent currently married, and 8 percent currently divorced, separated, or widowed (Luke 2006, 327–29). One wonders about the many who must be part-married—the bridewealth transfers and ceremonies only partly carried out. The numbers seem to obscure them. But it does seem that being married makes a bigger difference to women's sexual availability than to men's—something consistent with all else we know about Luo and western Kenyan marriage and mating.

How, besides romantic attraction, are women and girls induced or rewarded? Much like the Schwartz team's study, Luke's found that 72 percent of nonmarital sexual relationships involved transfers of cash, other valuables (gifts, meals, drinks, rent payments, rides, movie tickets), or both, usually from male to female, totaling about K. Sh. 430 (then about $6) per partnership in the previous month, which Luke estimated represented about 9 percent of men's mean monthly income. (It was, however, far less than the average takings of frankly "commercial" sex workers such as Schwartz et al., above, had reported from only a year or two before.). Some 51 percent was transferred in money, 27 percent in meals or drinks, 18 percent in (presumably durable) gifts, and the rest in rent payments and other things (Luke 2006, 329, 331).

These levels and forms of payment seem pretty consistent with reports of "survival sex" elsewhere in East Africa, as discussed earlier. A twist here is the high proportion of *un*married females (not just ones with failed or ended marriages) evidently reliant at least in part upon it. If it doesn't pay to be married, in Kisumu, it can and does pay to be mating; and for some it probably seems the easiest way to survive, let alone reproduce. In an environment of land shortage, job scarcity, high competition, and serious city- and region-wide economic insecurity, indeed, it may behoove women economically (whatever the moral implications) to think twice about pledging their troth to a man who may lose the ability or willingness to deliver the goods. (This is an implicit yet clear warning of both Abwunza 1997 and Francis 2000, on women's let-down predicaments and strategies for surmounting them in the lake basin countryside.) Instead, it can make economic sense to be able to shift from time to time to whichever palatable one happens to be currently solvent and giving.

As for exchange and condoms: Luke found that about 48 percent of the nonmarital sexual unions had involved condoms at last intercourse (2006, 328–29). As the amount of male-to-female transfers rose, she found consistently, the likelihood of condom use fell: for each K. Sh. 500 added, her regression suggested, the probability of condom use fell about 8 percent (pp. 338–39). But whether the transfers were made in cash or kind seemed to make no appreciable difference: "Nonmonetary transfers appear to be perfect substitutes for monetary compensation" as far as they concerned condom use (p. 339). Males gave younger females less than they gave older ones (K. Sh. 332 [or about U.S. $4.75] for the under-twenties as against K. Sh. 522 [or about U.S. $7.50] for the twenty-and-overs in a month). The younger females seemed not to be able to command as much as elder ones: maybe because their men knew these had fewer other options for earning. But in this study female age (whether above twenty or below) did not seem to matter much to condom use once male partner characteristics were controlled for (p. 342). Nor did male age or income seem to correlate well with condom use (though male education did increase it). What seemed to matter more was the sheer amount or flow of wealth transferred. It seemed to play a big part in governing not only whether a woman would participate in a sexual affair, but also with what sort of protection—and possibly, therefore, with how much trust, risk acceptance, or both.

What the Schwartz et al. study of the smaller towns, and the Luke study of the larger city, both indicate for western Kenya is a striking willingness of steady, well-acquainted partners, in relationships involving economic transfers, to drop their physical protection despite medical risks many of them acknowledge. This may be a gamble that some women, in particular, accept just in order to preserve *relationships they can count upon* to help keep them and any dependents fed, clothed, and cared for—as well as to satisfy any emotional needs they may experience. For young women, charged as they are and will be with so much of the child care and supervision, such continuity and reliability can matter more than they do for men, who, reproductively speaking, may do better by spreading their wealth and sowing their seeds more widely. Whether these strategies remain so effective or adaptive in an era of epidemic is more open to question.

Quite likely other factors also played a part without showing up on the survey forms. Probably some sex workers, clients, boyfriends were relying in part on diviners for predictions, or on prayer and sacrifice, direct mystical action (love magic), or a mix of these for protection—with or without physical prophylaxis too (see Wallman 1996, esp. Ch. 6; Whyte 1997; and more generally N. Schwartz 1989, 2000). In western Kenya as almost anywhere, it is not just other mortals, not just microbes, and not just manufactured devices that humans trust or distrust.

Child Parents, Parentless Children, and Parking People

Instances of unmarried pregnancy, wanted or not, raise hard questions about who owes whom what, in Kenya as anywhere. In impoverished times that make traditional bridewealth hard for young men to afford, and as new schooling, wage work, and "survival sex" work remove girls from home, cases of pregnancy without agreement of marriage seem to multiply. Pregnant schoolgirls and less fortunate youth highlight differences not just of sex but also of age, class, and status, challenging a court system more accessible to better-off and higher-schooled women and families than to others. In paternity suits, court officials have expended much effort trying to sort friendly gifts, "sugar daddy" payments, and the like from bridewealth—that is, to draw lines between dealings with marriage in mind and dealings without.

The gradual nature of bridewealth in "installments" makes the sorting a hard if interesting task. Historian Lynn Thomas, in Meru in central Kenya, has found young men wishing to put off paying bridewealth pitted against elder men wanting to preserve their daughters' and families' reputations by forcing such formal, "traditional" marriage. Women seeking to cement marriage tend to try to paint their ties as long-term, affectionate ones; men wanting out portray them as one-night stands or point to a woman's involvements with other men. Court cases hinge on the women's or girls' accounts of their histories of sexual encounters and their partners' gifts, which they are asked to recount in vivid detail. So much might be said almost anywhere. But these are things East African women and girls are ordinarily expected not to talk about outside intimate ties, especially not in front of their fathers—for in Meru as in Luo country and much of Kenya, ties with parents tend to be more distant and formal than with many other kin, for instance grandparents (and there may be added underlying concerns about incestual interest unspoken of). All this has put pregnant girls and young mothers in a very awkward and stressful social position (Thomas 2003, 122–34; see also Shadle 2003 on Gusii paternity suits).

Public policy on paternal obligations has flipped and flopped. A law of affiliation was passed under British administration in 1959, four years before independence, to offer pregnant unmarried women legal and economic redress, but it was repealed in the National Assembly amidst debate and some male derision in 1969. Some clearly deemed it a foreign cultural imposition. Researchers continued calling for new such measures decades later (for instance Onyango et al. 1991). Kenyans continue to debate whether to press paternity suits and how to handle them. That imported institutions like schools and magistrate courts have not expunged older ones like marriage dues, and maybe never will, leaves family law in an interesting but unstable condition.

These matters involve women's self-image, of course. As Marjorie Oludhe Macgoye (a British-born Kenyan Luo by marriage) suggests in her novel *Coming to Birth*, some Kenyan women take explicit "ownership" of their pregnancies and childbirth, deciding to view their own and their children's status (think of the stigma of "bastardy") as no longer dependent on the husbands' (and thus on parental negotiations, bridewealth, or ceremony), despite the pressures of patriarchy in surrounding society. Paulina, her rural-born heroine who endures miscarriage and anxiety, seeks thus to find her voice, her self confidence, in the city and in a man's world as her once powerful husband Martin shrinks in moral and political stature and goes astray. Procreation remains Paulina's fixation and hope—or is it trust?—but now not dependent procreation or wifehood. Macgoye depicts Paulina's transformation as a struggle and triumph in a way akin to national liberation from colonialism (Macgoye 2000; compare again Oculi 1968, Mwangi 1976).

Stirring indeed. Alas, though, there are those for whom it does not seem to work. For a single mother stretched with child-care duties, just as for a young nation, there remains the risk of dependency on patrons who may bully, condescend, and threaten to cut off aid at any time. Then it is up to friends, or up to grandparents possibly already overstretched with child care (at least in an era of HIV/AIDS). When funerals occur, the grandparents are so busy caring for guests that young mothers of only fourteen and fifteen can more easily be seen holding their own children (Cohen and Atieno

Odhiambo 1989, 97; see also Geissler et al. 2004 on grandparents, and Le-Vine et al. 1996 on same in Gusii country).

That single parents have not received adequate social support comes through when we look at some people who have turned to towns and cities for escape or survival. Here we approach direst need, this time from an urban vantage. The dissolution and refusal of marriage, the denial of paternity, the destitution of would-be dependents, and the reconstitution of quasi-families by children left largely to their own devices have lately been common phenomena in Kenya, arguably interlinked.

Obviously the methodological problems involved in collecting any survey information, let alone statistics, on street life are legion, just as for the intersecting topic of sex work. Here, though, are findings from two earnest efforts in that direction. The first is a major Kenyan-produced team study issued in 1991 surveying 634 destitute young people, "street" children and youths in six cities across Kenya, including Nairobi (56 percent of the sample) in central Kenya, Kisumu and Narok in western Kenya, and Kitui and Mombasa in eastern (Onyango et al. 1991). It also surveyed 32 of their parents and, for perspective, 80 members of the general public. The "street children" were actually aged six to twenty-six, a majority under sixteen years old, and actually not all were sleeping on the streets but working intermittently or roaming there, most already for two or three years (pp. iv, 10). Most were found to be migrants from other, mainly rural areas—not pulled as much as pushed, if only by felt poverty, perceived bad or renounced good will, or neglect. A second generation born in the cities was just appearing, and 2 percent of the sample were born on the streets (p. 18). About 91 percent male but joined by an increasing proportion of females, most came from single-parent "large" family homes (the average had six children), and most were currently living with groups of kin, friends, or both in the city, sharing food and whatever shelter they could make or find. Locally they were called "parking boys" (and lately "parking girls"), but that was far from all they did for a living. "Most of these children," wrote Onyango et al., "are involved in begging, pickpocketing, drug trafficking, child prostitution, scavenging, directing motorists to parking spaces and hawking" (p. 2).

This was their family situation in numbers, from some of the few sources available. About 78 percent confirmed having living parents (or presumably at least one). Some 30 percent lived with both parents in towns; 23 percent were living with their mothers alone as compared with 7 percent with their fathers alone, 11 percent with other relatives; and 18 percent with friends, 10 percent alone, and 4 percent in other ways (p. 14). As a whole, they could more easily specify their mothers' occupations than those of their fathers, from whom they generally seemed to feel more estranged. About 50 percent reported they were taking care of themselves; 19 percent cited their parents as a source of care, another 19 percent just their mothers, and only 4 percent just fathers (pp. 18–19). Large proportions, a majority evidently, were involved in dangerous activities like alcohol drinking and glue sniffing as well as at least intermittent commercial sex; and many of the girls in particular were victims of frequent physical if not also emotional abuse.

Kilbride et al. (2000) provide a more detailed portrait of the Nairobi street life scene, adding more on personal histories, families, and feelings as well as fresh numbers on AIDS orphans, refugees from interethnic clashes, and others. Most of the 400 "street children" they surveyed in and around 1994 were from central Kenya (53 percent were born in the slums of Nairobi or what have become so known) and only 4 percent were from Nyanza (proportions vary widely by neighborhood). In a sample of 98 parents of "street children," 62 percent said they, the parents, were not married and 78 percent said they were landless. The parents said the reason the latter were on the street was to look for money for themselves (76 percent) and their families (14 percent) (pp. 51–54). Forty-one percent regularly visited their families at home; another 25 percent occasionally, and 35 percent, some of these orphans, said never. Boys and girls both begged as a major survival strategy, and boys scavenged in refuse. Girls practiced survival sex but boys were not found involved in prostitution (p. 68). More than a few children formed husband and wife ties, in name or deed, with themselves and others (p. 79). They seemed to live more with friends than with siblings, but they used fictive kin terms often to familiarize friends (pp. 82–83). Luo children, more notably than others, imitated funerals in their play (p. 83–85).

In the main, the studies from Nairobi fairly closely join to others from around the lake. They show the landlessness or land shortage in rural areas, the immigration from the countryside, the fragmented families with tenuous threads sometimes still holding them together, the living in groups sharing food and shelter, and the dangerous sexual and other activities and frequent abuse. From these and other sources, it appears that many of the young romances and the urban youth mutual-support groups in Kenya, involving Luo, are inter- or multi-ethnic in background. Disadvantaged children and youth who have moved into cities, that is, seem often more cosmopolitan, less "tribalist," than adults in the rural communities the children come from (Lynsey Farrell, pers. comm. and my own impressions) or indeed many adults in the city. Macgoye's novelette of Nairobi's underside, *Street Life*, hopefully suggests they might not outgrow their intermixing as they age. The protagonist of this book, a typically Macgoye factual fiction, is a legless Luo street musician in the city with friends of varied ethnic stripes. Children and adults form associations when their families fail them. The message is that hope—for survival and for solidarity—never gets away unless one lets it.

Of course, sex for tangible rewards need have nothing to do with poverty. In and among these selfsame cities remains an ostentatiously wealthy Kenyan and East African elite. Its goings-on have hardly been surveyed, but they appear in the stories of authors like Macgoye, Mwangi, and Oculi and in op-ed pages. Its own families, commercial and political, sometimes dissolve and recombine not infrequently in Hollywood-style liaisons, often interethnic again, involving flesh if not also love secured by gifts and loans—of keys and cars, for instance. The opportunities for quick social mobility are big, and so too are the risks of malicious gossip, witchcraft accusations, and maybe witchcraft itself arranged by envious peers left behind.

If rich and poor can be profligate alike, town and country do not see each other as equally so. Rural dwellers, in western Kenya as in most places, speak of cities as dens of iniquity, casting suspicion on returned sojourners. Reflecting on a Nairobi neighborhood with many Luo, "The young of Siaya see Kaloleni as a place where sexuality is about money, something you pay for or are paid for, where the girls are spoiled: *okethore*" (Cohen and Atieno Odhiambo 1989: 98; see also Muga 1980, passim). Many urbanites with rural roots, as most have, like to send children back to farming homes for what they deem proper moral upbringing. They do so even though many small-town bars have sex workers and hourly rooms no less in evidence than do big city ones.

Nor do members of different ethno-linguistic groups look upon each other as equally accessible. In towns around the lake, Haya women from Bukoba, Tanzania, some high-status Ganda women from Uganda, and coastal Swahili women from Kenya and Tanzania gone inland have been stereotyped in the past, deservedly or not, for sexual license or prostitution (Southall and Gutkind 1980, 53–61; Macgoye 2000, 48). The reasons for these stereotypes might be many. Tagged are groups known for more general commercial orientation anyway (harder to bargain with, easier to bed?). It would be surprising if *no* people, places, or statuses got the reputation for being more lax or aggressive with their bodies, and anyway there are few places in East Africa, or maybe anywhere else, with mixed population where no commercial sex goes on, if only as a result of mismatching norms or opportunities in the gaps between.

The rewards, risks, and remedies of entrustment and betrayal involved in casual liaisons, quasi-marriage, and sex work — vital, emotive topics in times of straitened economy and major public health threats — await further study in country as well as city, and in the research frontier that is the small towns in between. These sites all connect; that is clear enough. Whatever moral judgments one casts upon unmarried unions, the one thing that both insiders and outsiders attribute as its biggest cause is poverty. Any real economic advancement for women that occurs can only strengthen their eventual social and legal position, including their bargaining power in cases of risk or exploitation. Or so one can hope.

The number of different forms of entrustment and obligation that an East African marriage can involve is impressive. While at the simplest, one group entrusts a woman to another, and the other turns over money and cattle in compensation for her and her progeny, the complexity of ritual forms and the contingencies of separation and divorce make the picture a lot more complex than this — let alone more complex than the simple "wasteful" or luxurious expenditure that banks or other financiers might perceive. The rules and expectations for a proper Luo marriage are so involved, and the number of "traditional" steps so many, that probably just about any real marriage is a matter of cutting some corners and making expedient compromises.

Like other East Africans, Luo people do not consider humans and animals, let alone humans and money, strictly commensurable in kind. They will tell you that no number of cattle can really repay the transfer of a woman, and no number of shillings can really substitute for a cow: at least, not for *one* of the cows, the one that makes a marriage a marriage. They often behave as if these things *were* interchangeable and will even, at times, use an idiom of borrowing and lending, even buying and selling — for marriage when it suits their purposes to do so. And yet these exchanges are never *just* loans, far less sales. The lapse of time involved, over years and even generations, is part of what prevents their becoming so.

There is another, subtler difference involving cumulative and non-collapsible reciprocities. When A brings B a cow, and B responds by giving A a goat or slaughtering it for both to eat, the one animal does not merely cancel out part of the first but also builds upon it by deepening or advancing a marriage performance in the expected sequence. And when a marriage ends in divorce, it cannot be merely assumed that the return of bridewealth animals simply annuls it, since this too is a further interaction that can require collective ritual, and since by this time there may be children involved who do not just go away.

Another aspect that cannot be captured in the idiom of loan or sale is the weblike interconnectedness of kin, and in other ways of friends too, in marital affairs. This is, in short, an area where terms like entrustment and obligation—deeper, less reductionist, more accommodating of third parties—fit much better.

Late twentieth-century scholarship on marriage in Africa made it clear that bridewealth and polygyny, like ethnicity, lineage, and clan, do not necessarily wilt and vanish with a shift from country to city, or with the growth of cities. Parkin's study found that by the late 1960s, Luo men had inflated bridewealth by using their urban incomes to invest in it. This, in his view, made them only more dependent on urban employment, but they used older cultural and semantic devices to maintain much control over women's marriages (1978, 319 and passim). What Parkin called a "dominant paradigm" "predicates family and lineage expansion through the exchange of women and bridewealth between exogamous descent groups" (p. 320). By this paradigm Luo structured not only their marriages but also the formation of ethnic-based urban associations and other responses to what Parkin found to be growing Kikuyu political and economic domination, and a shrinking share of the pie for Luo in the Kenyatta era.

Since that time the key players have changed and the pie shrunk further. In the Moi era, Kalenjins and their allies edged out both Luo and Gikuyu wealth and power, be they rural or urban based, and for a variety of reasons Kenya's economy and world position pretty steadily deteriorated. Gikuyu have made gains back under Moi's successor, the Gikuyu elder Mwai Kibaki; but whether Luo have is not so clear.

None of the political and economic turbulence Luo have lived through since Kenya's independence seems to have radically altered their basic understanding about marriage and the sexes, or eliminated customs like bridewealth that some might deem archaic at the turn of the millennium. The square dance of marriage continues to involve wide networks of kin, however simplified with ceremonial steps skipped (as indeed they may always have been). Bridewealth has changed form, incorporating new cash and long-distance payments, for instance, but it still takes animals, and the transfers still sometimes involve delays of decades if not generations. Wife givers continue to claim the upper hand over wife takers, and young married women keep getting called home when the cows come too slowly. If the entry of cash into bridewealth has, as some claim, made men think of women more as commodities (Buzzard 1982, 218), or as others state, rendered bridewealth less surely returnable, and thus damaged the credibility of affines (Parkin 1978, 110), cash and sales have not taken bridewealth over, and they are unlikely to do so soon. Where new laws and courts are involved in settling cases of company-keeping and "illegal" pregnancy, the parties find themselves arguing about whether gifts were crass enticements to sex, friendly charity, or failing attempts at bridewealth. Entrustment and obligation, in kinship and marriage, are likely to remain just as crucial to urban as to rural dwellers, and to the many (in the cities, probably the majority) who circulate between.

There is much scope for misunderstandings in the entrustments of marriage and other liaisons, not just between Luo and some foreigners but also between Luo men and Luo women, who may variously conceive of bridewealth machinations as sacred tradition or as self-interested male fictions and fantasies. The passage of time between transfer and counter-transfer makes amounts and values hard to compare and reckon. It throws equivalencies, qualitative or quantitative, open to question. Can a human's worth be measured in cows? Can one person borrow, lend, or pledge another? Are our bodies our own to lend or hire out at will? Whatever the answers, the calculation of entrustment and obligation extends to humans themselves, blurring any boundary line between the animate and inanimate, the economic and non-economic.

People in East Africa not only practice varied forms of marriage but also recognize different stages and degrees of marriedness. Upon drawing together these filaments on marriage and mating, and having followed the drift of people whose marriages or families have failed them and seen something of what they encounter when moving disadvantaged into the city or "the life," I must pass the reader along to other writers better versed on the particulars of urban sociology, epidemiology, or law. I do so, though, with a reminder about the wide variation in kinds of human union, and with four caveats.

First, what gets called prostitution or the "sex trade" is not just a matter of instant trade, cash on the barrel head, nor one of "selling" bodies or selves as so often framed. Much of it, maybe most, involves entrustment and obligation of one sort or another, materials included. If "the life" does not involve gifts given with a quid pro quo (of which sex may only be a one part of many)—and if it does not involve borrowed boots, beds, pills, or cars as its props—then it may well involve delayed payment. This last may or may not materialize, occasioning direct, indirect, or just imaginary reprisal in case of betrayal. Second, physical intimacy is not the same as emotional, though they can certainly coincide, whether there be payment involved or not. Third, neither the physical nor the emotional kind of intimacy can be measured merely in number of links in a chain of kinship. Biological and classificatory grandparents may be more intimate in their ways with grandchildren than either is with the intervening generation, depending on the matching or crossing of sexes and other factors. When the middle generation thins toward disappearing, the other two have more to do with each other than ever, but with limits on their logistical if not also emotional capacities.

Finally, a caveat on the kinds of information available and what not to assume about it. Words and numbers both have their uses, but different ones. *Figures must not be confused with facts, nor fictions with falsehoods.* To seasoned Africanists, the other pairing sometimes even seems more plausible; but let us settle on "no correlation." Truth has no mother discipline, nor any home genre in which to rest in comfort.

CHAPTER 7

Debts in Life and Death

Shared Responsibility and the Funerary Flow

In a lineage society there are in a sense only ancestors and children.
—JUDITH ABWUNZA

It has sometimes been observed that the living and the dead in tropical Africa are more present in each other's daily "lives" than in some other parts of the world. Ancestors, that is, are in a sense deemed not truly dead or gone; they have just changed form or been released from bodily attachment. In doing so they may not have lost power but gained it. Their spirits are seen to have needs, wants, and moods of their own, and ways of demanding attention. But they serve as guardians not only of custom and tradition but also of morality — of principles in adaptation to circumstance — and they are treated as intermediaries to divinity.

This chapter seeks to make a few points about the kinds of knowledge or belief that some deem basic truth and others superstition, and the kinds of entrustment, active trust with a material dimension, it can involve. Debt and obligation, in Luo country and elsewhere, are conceived to involve the dead as well as the living. These participate in each other's time and space — if these can be understood as time and space at all.[1] Luo and other humans extend into their spiritual lives some of the same understandings about credit and debt, and more broadly about entrustment and obligation, that they perceive in their affairs among themselves. The Luo reckoning of debt and obligation depends as much upon status as on deed. They are determined, that is, as much by who the parties were or are in the first place, and how, if at all, related, as by what they actually did. And yet Luo people perceive a sharing of responsibility among close kin in the consequences of spiritual, divine, or mystical action. In all of these elements of religious or cosmological ideology, there are elements of elasticity and causative pluralism that allow the elements of belief or knowledge to continue, not unquestioned (far from it), but never disproved; and the very questioning even seems part of what keeps them alive in Luo minds. Finally, I suggest, Luo understandings about the role of the dead in fiduciary life—or in the back and forth of economic and non-economic things—probably help peace among the living.

The Funerary Flow

Many Luo who have lived their lives in rural areas seldom if ever see a bigger gathering than a large funeral. Rarely is there one more lively, more gaudy, better nourishing, to politicians or public health workers more threatening, or to bureaucrats and old-school economists more frustrating, than a Luo mortuary rite. The funeral of an important man, for instance a senior statesman, may demonstrate for attendants the costumes, songs, dances, and poems that they deem the true and traditional Luo essence. Here is where one is most likely to see headdresses of ostrich feathers and necklaces of hippo tusks, to hear archaic poetic songs sung to the strains of the stringed *nyatiti*, and to hear stories of fallen heroes. Ceremonies for birth, naming, tooth extraction, graduation, and marriage may be skipped, but a funeral must not be skipped. Kinfolk and clan mates who work in towns and cities leave their jobs for a week to two at a time, risking not only their jobs but also, if they travel by road at night in Kenya, their own persons. In many ways, a funeral is the biggest celebration, maybe the biggest commitment—in short, about the biggest deal in Luo life.

Celebrations are expensive, and funerals, with their obligatory consumption of meat and other things, have been no exception. A proper Luo funeral requires contributions from dozens, maybe scores, of kin, friends, and neighbors; and while these are presented in a spirit of voluntarism, they are likely to be carefully recorded. For funerals are not only about life and death, but also about entrustment and obligation. It is not only the material and animal contributions but also the rides and transport, and the visitations and performances themselves that are remembered, repaid, and passed along. Needless to say, the Luo funeral cannot be covered here in a full or rounded way, but only the bare essentials connecting the economic and symbolic aspects can be treated.[2]

The economy of death in Luoland begins long before death itself and extends well past it. Many Luo rely exclusively on their lineage mates to cover the main expenses of their funerals, but some, particularly salaried ones like teachers, belong to burial societies that collect regular contributions as insurance against the expenses of members' death.[3] Since the locations of graves and abandoned homestead sites have become a main anchor of Luo identities, it is supremely important that burials happen at home. If one dies away from home, the body transport is far more expensive than for a living person, in part because the owner of the vehicle is likely to feel it necessary to hire an adept to hold a purification ceremony for the vehicle and those who rode in it or came into contact with the corpse.

The other big expense in funerals is animals for slaughter (some as sacrifices, some not), and other food for the gathering for several days, and commonly about two weeks for immediate kin and their spouses. Other visitors, more distantly related or perhaps delayed, may continue coming for many weeks or months—mobility and distances have only increased the period if anything—as individuals or in groups; and some kin, neighbors, or friends who have attended the funeral may return for further visits. Many of these come bearing gifts of cattle, sheep, or goats, or chickens, or grain and other grown foods by the sack—all of which are likely to be scrupulously recorded by some lettered adult in the home or family, in a lined school exercise book devoted to the purpose of ledger.

This is the stuff of both kinship and friendship. As David Goldenberg found in Karachuonyo, "Lineages linked only by the personal friendship of members often build up ties of obligation through participation in each other's funerals," and a delay may mean they feel compelled to make bigger contributions—for instance, providing rides, alcoholic beverages, a minstrel to play the nyatiti, or maybe a large gift on the hoof. "It is not unusual," as he notes, "for an urban worker to give nearly a month's salary toward the cost of a close kinsman's funeral" (Goldenberg 1982, 289, 297). Still, funeral visits are always symbolically significant, with or without substantial material or cash contributions, and the interlineage ties and obligations incurred may be remembered and reciprocated decades later, whether in material or symbolic ways.

The reciprocities of funeral visitations help explain why Luo often seem to feel funeral attendance obligatory, even for kin or affines at several degrees of remove. They do not explain it all, since Luo-speakers, like people everywhere, are quite capable of understanding the concept of proxies for groups. But the nature of the gathering helps explain the rest, since funerals, like market gatherings, are occasions where countless social activities go on.

When a large van full of relatives departs for the funeral of a distant lineage or clan, it is not hard to sense not just the sadness but also the excitement, and seemingly opposite emotions sometimes connect like the ends of a circle.[4] And some of the poorest, as noted, depend heavily on funerals and other large gatherings for meat and basic alimentation, just as ritual specialists, carpenters, musicians, and others depend partly on funerals for their living too.

A Luo Funeral

The case of the funeral of one of my neighbors in Kanyamkago, a plantation worker in his late thirties who died suddenly in a road accident in the early 1990s, illustrates some of the entrustments and obligations surrounding funerals and the feeding of their guests. The following brief account of some of the early stages of the funeral shows how economic decisions blend with symbolic ones.

Since the police had sent the man's body to Homa Bay for a postmortem, his family in Kanyamkago dispatched some of his cousins to arrange its transport home from there. They first approached his former employer, a factory owner, and asked his help. He obliged with the loan of a truck, a coffin, and a cash contribution of Shs. 2,200 (then about U.S. $92). Whether he did this out of a sense of indebtedness for past work or just patronly friendship I was unable to ascertain. Meanwhile, relatives and others had begun arriving at the home of the deceased, and the family slaughtered a ram to feed them. Its skin, not needed for any particular ritual but still valuable and symbolically potent, was sold and the money used to buy maize flour and paraffin for the entire funeral—a convenient way of avoiding a decision about whom to give the ram to. (It would ordinarily have been given to an honored in-law, who would then likely feel morally obliged to pass it on.)

When, two days after death, the body arrived home, kin and church mates (and some both) had dug the grave outside the man's house (that is, his wife's) but within the homestead, as current custom prescribed. The family did not want it buried, however, until the assistant chief, a close friend of the dead man, arrived. So they left the body in the coffin in the grave, uncovered, overnight; and when the assistant chief showed in the morning, the clan held a fundraiser with him and the county councilor's wife as guests of honor. Many guests had arrived by this time, and the family slaughtered a bull to feed them. As with the ram, they aimed by this as much to feed the living crowd as to feed the ancestors. A family sponsoring a funeral satisfies the dead, some say, precisely by feeding the guests well, among other things. Here, as so often, the role of the third party in an exchange is vital to its meaning.

At the graveside, before the grave was filled in, a classificatory father read a *neno* (lit., "word"), a life history eulogy for the dead man. An unrelated neighbor also showed up at the graveside to make a claim for a debt of four *debe* tins of maize worth some Shs. 600, asking, before witnesses, that the surviving agnatic kin find a way to pay it off. While the lineage representatives present willingly agreed that this would be done, the man's claim gave rise to mixed responses from other guests at the funeral. While most thought him quite within his rights, some thought his insistence on recovering the loan rather tight-fisted. (Weeks later, no one yet knew how the loan would be repaid, but it was supposed that the lineage, *anyuola*, would be asked for contributions at the end of the funeral proceedings.)

On the third day, in the early morning, the dead man's clansmen launched the *tero buru* (or just *buru*, lit. ashes) ceremony. This rite might, in another case, have been left until later—in the case of the death of an old man with no young widows, for instance, it might be delayed to give more guests time to come. But since, in Luo country, no bereaved woman may have sex before the buru or risk bringing chira to the home and lineage, and since there were young women in the family, some of the elders felt that it was prudent to get the buru out of the way early, lest anyone might be tempted soon into sex before it.

Following current custom, the buru involved a party of young men's hiking defiantly with spears to the edge of a neighboring oganda, with a rooster and glowing embers (or, now, at least a matchbook) from home as if to colonize wilderness. A close agnate of the dead man ritually slaughtered the bird as a sacrifice (*misango*), dripping the blood from its neck on the ground for the ancestors to eat and roasting the bird on the spot for the men to pull apart and consume in the bush. A subgroup set off to collect reeds from the lake to show they had been that far. Formerly, the highly symbolic buru ceremony—using the same symbols of spear, rooster, and fire used in founding a new homestead—sometimes started a battle; and still now, people of the encroached territory may resist the intruders' sacrifice, so these usually carry it out quickly and sometimes even surreptitiously on arrival.[5] Emin Ochieng' Opere summed things up in an interview at the time of the buru: "It is our belief that when the blood sprinkles or drops on the ground, the ancestors are already there waiting to receive it. The blood of the chicken [rooster] is feeding the ancestors, and the demons (*jochiende*) are driven away if at all they are angry. But the good ones remain. You eat the meat. Just the way warriors used to do, without *ugali* (*kuon*)."

The "war" party triumphantly returned home later that day, met along the way by an exuberant throng from home in ritual regalia complete with spears, headdresses, musical instruments made from kudu horns, and so on. Back at the homestead, the assembled crowd was divided into groups and seated by *ogendni* (clan federations) of origin.

But soon the host lineage had a problem. It was short of food for the guests. There were no cattle to kill, the family having already slaughtered all it could immediately call in. The dead man's father's brother offered a ram, and this was slaughtered. This was just a stopgap measure, though; lineage members conceded that it ought to have been a bull, and that they would have to invite the guests back another day when a bull was available, not just to make good a debt (anyway, the meat was reckoned to count against contributions the guests had brought) but also to satisfy tradition and the dead. As one cousin of the deceased man put it, "If no one had at least a ram, you have to buy. The clan has to buy. But not to go without something. If it goes without something, then he will be annoyed, he will say that 'my colleagues came here,' he will appear again as *jachien* [angry ghost], that one will bring *tipo* [ghost(s) or spirit(s)]. So all these are done in order to clear away the *tipo*."

One of my informants also spoke of the ram as a "bull," in much the same way that Evans-Pritchard's Luo informants spoke of a sacrificed wild cucumber, another makeshift substitute, as an "ox." The Luo informants from Sakwa *oganda* helped resolve a question that countless incredulous students have asked of Evans-Pritchard's work: "Do they really think this can substitute for a bull/ox?" The answer, in this case, was yes and no. The ram could suffice to patch over the void in the ritual, but it could not fully satisfy the obligation the hosts owed to the guests, or in turn, to the spirits of the dead who desired their satisfaction. Otherwise there would have been no need for the guests to return.

Other funeral stages, like the semiprivate ceremony of head-shaving (*liedo*) for immediate kin by an elderly woman herbalist (to mark these mourners and symbolize purification and a fresh beginning), involve their own sacrifices and feasts, as well as gift-payments to the specialists who carry them out. For this event local beer was once a standard libation, but purchased tea had become more common by the early 1990s. The ceremony of *yawo dhoot* (opening the door), for out-married women who come back to revisit their home, is another of the occasions for feasting. Whereas men more commonly are expected to take live animals or meat to present when making ceremonial visits, women doing so more usually take vegetable foods, as if to symbolize in an exaggerated way their respective productive roles in society. This occasion I would have to miss.

In the three weeks following the death and burial, as a part of the series of continuing ceremonies, more animals and food would come in with visitors paying their respects, and loans would be called in. The men of the homestead would slaughter four more bulls to honor visitors who came from different ogendni, arriving, some in groups, to pay their respects. The beasts would be divided, cooked, and eaten, and their skins sent home with the relatives being honored on their respective occasions.[6] When I left the country at the end of those three weeks, visitors were still coming.

Debts beyond the Grave

Luo debts are heritable; as noted earlier, they are not automatically cancelled upon an individual's death. This is true not just of the longer-term and more sacred debts like bridewealth (which in any case is not an individual affair but an interfamilial one) but also of smaller, quotidian ones. As Luo speak of it, it is the right of a creditor to announce or disclaim a debt—be it in money, food, or animals—at a funeral. The reckoning must be announced before the dead may be buried. Whether the claimant will forgive the debt will depend on how closely related he (or she) is to the dead, on what is owed, and on how he happens to feel about it or is persuaded by others to act.

Where animal debts are concerned, a borrower deemed unrelated to a lender is usually expected to return the animals during his (or her) lifetime. In this lending, that is, the "generalized reciprocity" (in Marshall Sahlins's term [1974, 199]) is expected to follow the closer kin ties; as kinship attenuates the reciprocities begin to look more "balanced" and less likely to be forgiven. Witnesses help safeguard claims, and the lenders may resort to chiefs and the courts—if they can afford the expenses—in case of failure to return the animals.

Most creditors of money and food debts seem to forgive them at funerals, a point on which Goldenberg's impression (1982, 286) agrees with my own. This is, of course, an opportunity for the creditor to gain prestige in a display of magnanimity or compassion—even the mere announcing that one has been a creditor gains one something of the direct respect that wealthy people enjoy in Luo country—and the net economic effect of the "speak-offs" is likely to be redistributive. But the creditor need not forgive the debtor, and the most important thing is that it be clarified before witnesses whether each debt is forgiven or carried over to any survivors.

One can be born, as all of this implies, with debts to pay or to collect. In a sense, everyone is born with a debt to one's mother's natal kin, since bridewealth does not fully discharge the debt for a woman exchanged in marriage but only legitimizes the bond and the birth of the offspring. If nothing else, one owes them continuing respect. Heritable debt is the concept reflected in a Luo origin tale about dog and human. In brief, the primeval pair of dogs, a married pair, were being chased by Opul the hyena, who had already greedily eaten their pups one by one, attempted to eat the dog wife, and after the husband arrived and joined the fray, nearly defeated the dogs in a struggle. They found refuge in man's house, and man listened with sympathy. Dog, eternally grateful, lived with man ever after (Onyango-Ogutu and Roscoe 1981, 45–46).

However alien the heritability of debts might appear to members of societies that fancy themselves more individualistic, they are part and parcel of an East African way of construing personhood. A father and a son are not entirely separate people—nor are a mother and daughter—but instead, one is deemed something like an aspect or appendage of the other (one's sons and descendants are one's legs, as some speak of it). This is one reason why a dead man may marry, and why an offense against a person is also one against a lineage or clan.

Widow Inheritance and Ghost Marriage

Luo, like Nuer and other Nilotic people, have long practiced what get called widow inheritance, ghost marriage, and a particular form of woman–woman marriage—institutions (and names for them) made famous in the Nuer case by Evans-Pritchard (1951). These practices have persisted despite the efforts of Christian churchmen and -women to end them, and despite the campaigns of public health workers to stamp out widow inheritance as an HIV hazard. For they connect to much else in Luo and broader Nilotic culture. Rather than attempt a full description and explanation of each of these customs, which involve much discussion and ritual, we may simply note here that they can involve entrustments, obligations, and other economic and quasi-economic transactions in which the dead play a part.

Death, to Luo, need not spell the end of a marriage. A man's name and position in a lineage outlive him, and so does his wife's "house" (*ot*) as the focal point of a potential new lineage segment: the basis of the memories, mixed with dream apparitions, imaginings, prayers, and sacrifices, that will keep one "alive" after death. So they are made to carry on their own economic and symbolic dealings after husband or wife is dead and buried.

A woman sent to live and procreate with a man and his kin in return for bridewealth who then dies leaves a debt to the wife-takers unfulfilled. In the past, her natal kin have been expected to make this good by sending another woman, a sister or cousin, in her stead, to "fill the basket" (*pong'o dhuong'o*) as an adopted or "planted daughter" (*nyar pidho*): something that could take many years to happen. The widower was expected to pay bridewealth again, but at a "discount," evidently because the cows and other wealth needed to negotiate and establish an affinal relationship (as distinct from maintaining one) were no longer needed.[7] A man whose wife was "barren" was entitled, according to Wilson (1968, 134), to demand a replacement, and to delay starting the required bridewealth transfers for her.[8]

What if a husband dies first? The new widow enters, or is pressed into, a union (here called *ter*) with a kinsman of her dead husband, preferably a full or classificatory brother, as her *jater*, replacement mate. *Ter* gets translated various ways, including widow inheritance, guardianship, or just takeover. It is also sometimes called the levirate, but it is not precisely the same as leviratic "marriage" or the mirror image of the "sororate" above (since some of the Luo nuptial ceremonies, *meko*, are skipped this time around). After being joined by her jater the woman continues to bear any children to the name of the dead man; that is, they will grow up regarded as belonging to him and his lineage, and so do their own. The widow's natal kin do not require further bridewealth for the second union; the bridewealth/childwealth that has been, or is still being, "paid" for the first carries over and guarantees the husband's lineage its claim over her productive and reproductive powers, and over those offspring.[9] All this shows clearly how partners in marriage have been understood to be filling roles within broader networks of kin.

Some widows have, in recent decades, resisted being "inherited" on the grounds that they want to make their own choice. Their submitting to a takeover effectively means relinquishing titular control over land, so women's struggle for land rights (among those who do struggle) is all tied up with widow inheritance. Public health workers concerned about sexually transmitted diseases have targeted "widow inheritance" (or "levirate," as some loosely call it) for "reeducation" campaigns, suspecting this to be one of the main ways HIV and other viruses spread. But the custom continues anyway, in part, it seems, because some women continue to feel inadequately protected from other men's predations if left on their own.

Women whose husbands have died without leaving children are sometimes known to "marry" other women to the names—that is the ghosts and positions—of their deceased husbands. For the bridewealth they may use animals from the husband's herd. Women who are widowed with daughters, but no sons, may use the bridewealth from those daughters' marriages for this purpose (Wilson 1968, 127), thus passing the animals from ownership by the living to ostensible ownership by the dead, at least until they are passed along. The new bride, *chi mandu* (wife [of the] wealth), is expected to cohabit with one of the dead man's agnatic kin. Such a man thus becomes the biological *genitor*, but not the social *pater*, of children she bears.[10] The widow who paid the bridewealth becomes that. Ghost marriage thus defies any boundaries not only between one species and another (woman for cattle) but also between the living and the dead, and between one sex and another. These are the lengths to which Luo have gone in the past to ensure that every man leave progeny. Continuity is the key.

Compensating for Violence

Threats to life, and to lineage continuity, come from other people and their ill will. The foregoing pages have described something of how Luo respond to death, and of what part entrustment and obligation play in relating the living to the dead and in helping family overcome mortality. But there are different ways of dying, and some entail special dealings. Where one has killed another, the question arises whether the death has caused a debt, and if so, how it might be settled.

It is difficult now to learn the details of homicide compensations of times past, before the British-imposed courts took over in the first decades of the twentieth century. But early British colonial officers' ethnographic sketches occasionally mention blood-payment obligations among Luo. A report from D. R. Crampton, Kisumu District commissioner in 1909, one of the earliest available on the topic, reads, "If committed against a member of the same clan, the family of the deceased would receive from five to ten head of cattle according to the arrangement made by the elders at the funeral of the deceased man. . . . In the case of a man of a neighboring clan being killed it is possible that if the two clans were on friendly terms with one another that a settlement would be paid or a girl handed over to the uncle of [the] deceased man. More probably the deceased man's relatives would take matters into their own hands and endeavour to obtain revenge. . . . Murder of women is practically unknown except the murder of a wife by her husband. . . . In this case no punishment would be meted out to the murderer who would be regarded as having been possessed by an evil spirit in that he had killed his wife, i.e. had destroyed his own valuable property."[11]

Amounts demanded in homicide payments apparently depended on kinship distance between the parties, as another officer noted in about 1934: "In the case of the murder of a man of another clan, compensation . . . would either take the forms of a payment of sufficient cattle to buy a wife, or of a girl to be taken to wife by a male relative of the deceased. If a near relation was killed, an ox was given to the Elders, and another ox slaughtered and eaten by all members of the family."[12] What is hard to know is just what these officers meant by "clan," since many used it for what anthropologists call a lineage too. But the general point is clear enough. Taking a life created a debt, and what you owed your victim's people depended on how closely you were related.

Comparable and more detailed findings, linking the nature, likelihood, and amounts of homicide settlements with kinship distance, were obtained among the Nuer (Evans-Pritchard 1940, 150–62), the Shilluk (Howell 1952), and other Nilotic groups of the Sudan.[13] Evans-Pritchard found among the Nuer that "In theory forty to fifty head of cattle are paid, but it is unlikely that they will all be paid at once and the debt may continue for years" (1940, 152). In their long protraction and in their involvement of patrilineal kin, homicide debts among Nilotes up to early colonial times resembled bridewealth debts up to the present.

The nature of homicide debts, compensations, and vengeance also depended on the status of the victim and the accused with respect to the land and the lineages on it—that is, on who belonged there and how. The fundamental social inequality noted earlier between *weg lowo* (masters of the land) and *jodak* (land clients or visitors) came into full play after a killing. Anthropologist Gordon Wilson, codifier of Luo land law for the colonial government, generalized in retrospect from cases, as the colonial years drew to a close (1960, 183): "Those living within a clan or tribal area who are not members of the clan or tribe [here meaning oganda] are regarded as squatters (*jodak*) with limited rights. . . . Murder committed by such persons invokes immediate counteraction on the part of those who belong to the land by virtue of descent, *jogweng*. Cattle could be taken by force, the culprit banished or killed, and/or a woman from that group married by a member of the dead man's minimal lineage, his *jokakwaro*, to raise seed unto the dead man's name. This woman would be regarded as the wife of the dead man, her children would placate his ghost as father (*wuoro*), the physical father would serve merely as genitor."[14]

While a woman for a man, a person for a person, might look or feel like strict reciprocity in some symbolic sense, it is not in any practical sense. For one thing, male and female lives have at least until lately been accorded different value, as seen earlier. For another, her children (even if deemed the dead man's) may multiply indefinitely. Ghosts with many offspring are more benevolent, and easier to appease, than those without: in a Luo understanding, they are richer in what matters. All this means that a homicide "settlement" is likely to be little other than a convenient fiction, useful for making peace but not a true representation of, or substitution for, what has been taken or given.

Sometimes, ironically, the way to get even is not to kill the killer, but to embrace him. In a variation Wilson documented (1960, 183–84; cf. Whisson 1961), a jadak from Ugenya oganda living among the people of the Seme oganda (then in Central Nyanza, now in Kisumu District), killed the son of a Seme elder. A war broke out between Seme and Ugenya. Eventually, by mutual agreement between the elders of the two ogendni, the killer was forced to marry the daughter of the deceased's father, to move to her home (contrary to ordinary Luo virilocal practice) and to become absorbed as a man of Seme. Their descendants took her name, rather than his, as their clan eponym (Kia, in Kakia). In this way his original social identity was buried for good. Seme had reclaimed a son, and a link in its eventual genealogy, for the one lost.

In murder as in other infractions of custom, punishments have long been more severe for jodak than for weg lowo.[15] Debts are acquired not just actively but also passively, and the nature of debts and their compensation depends as much on who you are—that is, to whom you are connected—as on what you have done. This is a message that keeps appearing in Luo entrustments and obligations of many kinds.

Luo responses to a killing have in the past included both purification and/or banishment for the killer (since a death is tainting for the killer and others in contact with the dead) and compensation for the victim. Angry

ghosts (*jochiende*) of persons killed or otherwise wronged participate in their
own vengeance, and some Luo say they can travel to the killer's home to
carry it out.[16] For killings done in momentary anger, punishment not meted
by one agency gets meted by another: God, the ancestors, or the forces of
nature will see to it that a killer is punished, if humans by themselves do not
(Wilson 1960, 183). It is as if everyone had a quota.

As the instances of bridewealth and homicide payments show, some debts
have sacred as well as economic dimensions. Settlement of debt disputes in
and around Luoland has commonly involved oath-taking. One form recalled
or imagined in 1933 from the unspecified "olden days" occurred "when quar-
rels took place over debts and no settlement could be arrived at." It consisted
of "spearing the *mtupa* tree (*Euphorbia candelabra*) with a spear which had
been blooded with human blood and swearing to abstain from any particular act." The most powerful oath of all, reported by the same writer, was
practiced between members of different ogendni; it was sworn by the two
parties while an officiating medicine man cut in half a dog whose head was
held by one side and hind legs by the other.[17] Luo informants today still tell
of such oaths, and some still practice others, sometimes invoking ancestors
or divinity; a simple one involves pounding the ground with a stick while
swearing or cursing. While the reconstructions and interpretations of these
past and present practices are always debatable, the most important debts in
Luo life have never been just secular matters.

National government, in Kenya and neighboring countries, has all but
taken over the prerogative of punishment, for much of recent history. It has
done so not just through the courts and jails but also in places hard to reach
or out of political favor, by paramilitary police invasions. But whenever this
and other governments fail or wane in their local presence, local mechanisms
such as vigilante groups are likely to arise to take matters into their own
hands (Heald 1998, Abrahams 1998). And spirits are still deemed to concern
themselves with lesser offenses and slights than the law picks up, or ones
which, like neglect of an ancestor, do not appear in common law statutes.

Spirits and Shared Responsibility

The large literature on bridewealth and on homicide compensations
in Nilotic societies highlights the shared responsibility of some kinds of kin
for the actions of others. If, in decades past, a Nuer man speared a distant
clansman, for instance, not just the spearer but also his brother may be ex-

pected to pay cattle or other compensation or be speared in return (Evans-Pritchard 1940). Much the same seems to have been true of Luo.

We have already seen that debts in Luo country are heritable; sons are responsible for the borrowings of their fathers. Luo daughters in the past could be handed over as compensation for homicide, and still, in recent decades, as pledges during famines or other crises. Luo understandings about *pesa makech*, bitter money, also rest on assumptions about shared responsibility. Sell gold or patrimonial land for a selfish reason, and the one whose cattle catch diseases may be not you but your brother, or your children or grandchildren. The onus is shared both laterally and lineally.

This is all part of a recurring pattern in tropical Africa: the sharing of responsibility for debts or credits—or help or harm, or good or bad deeds—among certain close kin. Which kin depends on culture and context, and the answer need not follow simple rules like patriliny or matriliny.[18]

Certainly in the clustering or individuation of allocated responsibility for deeds and misdeeds, cultures vary enormously.[19] While some non-Africans (and indeed some Africans) may consider an assumption of shared responsibility like that of Luo to be madness, yet there is method in it. Three kinds come to mind. First, the assumption may well help ensure adherence to cultural norms by giving people incentive to monitor each other's actions.[20] Second, it may help spread around the material burdens and rewards of compensation payments, thus helping even out the endowments of homes and farms. Finally, it may help perpetuate the very belief or understanding in question about mystical causes and effects, since wherever misfortune eventually *does* befall someone of a family, household, lineage, or other set of persons, its cause may be attributed to the known misdeeds of some kin or forebear. The sharing, that is, serves as a kind of elastic, allowing the belief/knowledge the leeway to prove itself right over time.

To sum up, Luo concepts of debt and obligation reach beyond the world of living humans to involve also the world of spirits and divinity. A man or woman who dies with anger or a grudge against kin for fair treatment is likely to return as a malevolent spirit or ghost (*jachien,* pl. *jochiende;* also sometimes called *tipo*). The causes perceived can be many; they include the failure of the kin to have provided cattle for bridewealth (in the case of a man) or a proper marriage or funeral ceremony, or failure to avenge a murder or exact compensation for it (Abe 1978). Jochiende manifest themselves in dreams or hallucinations or by causing compulsive behavior in persons possessed. Alternatively they show up as fleeting human or animal forms in daylight, or their presence is inferred from otherwise inexplicable misfortunes like disappearing cattle or a house burning. The spirit may avenge not only the one who neglected the living but also his or her children or grandchildren, or seemingly innocent others to whom a curse has been transferred in a cleansing ritual (*dilo*). A belief like this one, insofar as it has a function, may help ensure that the living share responsibilities for preventing misdeeds by their kin or neighbors. It is not just individuals who are responsible for settling up their obligations fairly, but others in their daily lives too.

This chapter and the preceding one have, I hope, given some idea of how Luo and some other related people conceive of their entrustments and obligations—surrounding marriage, funerals, homicide, and the kinds of injuries and damages lawyers call "torts"—to crisscross the boundary between life and death. The living, that is, conceive of their fiduciary dealings with each other in terms that include the dead and unborn. The importance of "vertical" (intergenerational) kinship, the close involvement of humans with domesticated animals where reproductive matters are concerned, the sharing of presumed responsibilities, and the watchful presence of third parties seen or unseen are all important themes here. In pages that follow we look more closely at inheritance and then at sacrifices and offerings, thoughts and acts more often called religious, involving not just ancestral and other spirits but also divinity of a more ultimate, more encompassing sort.

CHAPTER 8

In the Passing

Inheritance of Things and Persons

We were talking about how an old *mzee* [senior man] can distribute his wealth to
his family members, especially when he's dead.
—OKELLO MIDINGO, A MIDDLE-AGED LUO FROM SAKWA

Inheritance provides a window into some big ideas. Who passes
what to whom communicates messages about relative worth, not
just of "property" but also of people, and of the bonds between them. It
illustrates disparities in the way the sexes and ages are trusted and treated
and makes concrete a people's ideas about the importance of the dead and
unborn in the lives of the living. Looked at over generations, inheritance is
serial entrustment. Of what we receive, we pass along.

In Africa and elsewhere there are societies, notably ones that rely heavily
on mobile foraging, where inheritance is altogether downplayed, and where
most of a person's possessions are destroyed upon his or her death. Such a
people the Luo are not.

Luo inheritance defies simple rules. It is mainly about men's wealth,
since men claim most wealth to begin with; but even so, some of the most
important things ostensibly inherited by men are used mainly by women,
and women also bequeath and receive directly. Men make decisions individually, but in consultation with groups that may make them rethink their
decisions. Inheritance mainly follows patrilineal paths, but there is room in
it for choosing and playing favorites, both within and outside a patrilineage.
A will expressed in life often takes effect only after death, and after witnesses
may have gone senile or died themselves. Inheritance among Luo is partible
(as among most eastern Africans), but in it there are elements of both primo-
geniture and ultimogeniture—that is, favoring the first and the last born (or
married). It is egalitarian, but usually unequal. The pattern of inheritance de-

pends on what it is that is being inherited, as culturally significant items have their own rules, and some get allocated only temporarily. It is not only goods and services that Luo inherit, but also bads and disservices. More broadly, not just these things but also human beings themselves, widows and other humans, are taken over and are sometimes spoken of as being inherited. The same idiom, that is, gets applied to beings subject to very different kinds of ownership or control. Let us spell out these points.[1]

First, though, to clarify some terms. It is worthwhile distinguishing between "inheritance" (transfer after a death) and "devolution *inter vivos*" (transfer down the generations between the living), although inheritance depends on living survivors to implement it. By "heritage" I refer to a kind of inheritance in which a large group—for instance, the Luo as a whole—supposedly share an interest. Some kinds of heritage—for instance, children's songs or games passed along with no adult help—may continue indefinitely through time without involving death. Devolution, inheritance, and heritage are terms that may refer to the conveyance of things without material substance, such as the passing of songs, dances, or healing practices. Such stewarded incorporeal legacies are found in Luo family and clan traditions (just as they are found among, say, native people of the northwest coast of North America, who treat them as important forms of private property, or among software firms that copyright bits and bytes). But the discussion that follows, in keeping with the more restricted definition of entrustment used for this study, concentrates mainly on things with material dimensions. In common English parlance, "sentimental value" is sometimes assumed to refer to tokens or keepsakes, things of small size or little material use value, such as a ring or hairpin. But large and useful things like land or livestock can also have sentimental value. In Luo country these have plenty—just as do the framed photos of glowering elders and ancestors that some Luo in recent decades have placed so conspicuously high and alone on interior walls, keeping watch over the living.

Deciding How to Divvy: A Prayer and a Promise

Luo expect a man to make his intentions known well before he is at death's door, and some prefer that he set the process in motion while still healthy and strong. An aging Luo man may vet his intentions with a trusted friend before announcing them to his lineage (*anyuola*) mates; and it is no surprise to anyone if the friend persuades him to rethink his plans. The meeting then with the clansmen is called *pogo-lemo,* dividing prayer, or simply *lemo,* prayer. Partition and prayer have in common the concept of expressing a wish—and a share of an inheritance is also known as lemo as if to emphasize the will behind the thing. Lemo also suggests that it is not directly within the power of the one bequeathing to ensure what will happen after he dies, but the point is a debatable one, since one who flouts the will of a dead person is said to bring a curse down upon him- or herself.

In addition to announcing his will, in the pogo-lemo, a man also lays out before his kinsmen what debts and credits he holds with regard to other people outside his family or lineage. They know that if he dies before these are settled they will inherit them, minus any that creditors who show up at the funeral might forgive.

The main protagonist in a pogo-lemo is usually a man, but women are at issue too. One man I spoke to in Kanyamkago said he considered it normal for a man to inform his sons of his plans for inheritance before his wives. Few women seem to hold their own pogo-lemo gatherings—either because they feel they have little to bestow, or because they deem this ceremony a male prerogative anyway. A man is presumed to know and understand the different possessions and capabilities of his wives and offspring before deciding on what parts of his land and property should pass to them, and to take these factors into account. Thus a wife with four sons and no daughters can expect to receive a bigger share of inheritance than one with only two sons and three daughters, since in the former case, the sons have no sisters to bring in bridewealth for them to use in their own marriages, and in the latter, they have more than their number to do it.

But something big intervenes: the principle of seniority. Men expect, rightly or wrongly, that first wives will object to junior wives' receiving amounts equal to their own. Since most first marriages are made at a time in the life cycle when men control very little wealth, men expect their first wives to look upon all subsequently acquired wealth, which for certain purposes includes the "following" wives themselves, and the bridewealth that procured them, as something the first wives too can claim to have achieved or managed in the adventure of married life. First wives, say their menfolk, consider themselves and their labor the very sources of later wives as far as the homestead is concerned: these are all a product of their initial partnership and their joint efforts.

Such claims about first wives fit the standard image of the domineering first wife who orders the second about. This is a common theme in Luo lore

if not also in day-to-day lived experience. I have more than once heard Luo
men say that they would like to divide their inheritance equally among co-
wives, but feared to do it because of what their first wives might do. Those
women's actual sentiments are, I believe, more varied, but that is a men's
stereotype. Often enough a junior wife out-reproduces a senior, or has more
sons; but if my best informants are correct, it is not common for her and her
offspring to receive more than the first wife and hers. In interviews with Luo
and Luo-Suba elders from different locations about inheritance, this tension
between needs and seniority came up time and again, and while most observed that the first wife of two or even three usually ends up with two-thirds
or even three-quarters, if there are two co-wives—the leopard's if not lion's
share—no one pretended that this solves the problem.

Complicating the picture further is that sons are not merely passive recipients of inheritance, however much the gerontocratic ideal would have
them so. They can and do express complaint, and when they deem their
fathers to have neglected their bridewealth needs, they can occasionally become violent: an abomination in the Luo order of things.[2] If this latent threat
is not enough, there is another danger. If one of them should die unmarried, for want of bridewealth that a father could have allocated him but did
not, his spirit (*tipo*) is likely to become an angry, vengeful ghost (*jachien*).
Now since a jachien does not necessarily afflict the one who perpetrated the
grievance but might well strike someone related but otherwise innocent, like
a small child, this structure of beliefs encourages other family members to
exert pressure on the husband to see to his children's needs and make sure
no one gets left out. Luo speak of marriage as something like a basic human
right, and whoever willfully deprives one of it, or stands by to let that happen, may be in for it later, or see illness or suffering close at hand.

Every son, Luo will say, should receive at least one head of cattle. There is
a world of difference between getting something and getting nothing. Symbolically there is more difference between one cow and none than between
one and three. The principle is familiar from marriage: it will be recalled
that a marriage with no cow transferred is deemed no marriage, though other
wealth be transferred in plenty. Since, in a society allowing polygamy, it often
happens that some half-brothers have grown to adult and marriageable age
while others are still in their infancy, one who is allocating shares of inheritance must weigh present against future needs. This is no easy task, since an
animal or two assigned to an infant son may, by the time he has grown up,
parent a small herd or may simply sicken and die. One can only guess, hope,
and pray.

A case from my own experience illustrates a process of allocating animals to the young. In 1991, to thank a family in Kanyamkago who had hosted me for a sojourn, I bought a heifer and presented her to them, suggesting that she not be slaughtered on my behalf but kept for milking and maybe breeding. I did not necessarily intend that the animal would be assigned to an individual, or at least not right away, but chose a female animal precisely so that more than one could share soon in the benefit. But the *wuon dala*, a man in his sixties, took it upon himself to assign the young cow immediately to one son's future bridewealth, and other family members acquiesced in this. He chose (at least for the time being) the youngest son (then aged twelve) of the *mikayi*, the senior co-wife, on whose side of the *dala* I had been staying. In doing so, he passed over the nearer-term interests of two elder, unmarried sons of the same woman, and one toddler son of the junior co-wife. This toddler would be the one, as the most junior, expected most likely to receive the father's *mondo*, his personal holdings left over after the main inheritance were done—the charitable element of ultimogeniture in the Luo way, for the one whose position was presumed most dependent and vulnerable. But had the father chosen that youngest son—that is, the youngest of the half-brothers—for the heifer, then the ones on "our side," the mikayi's side, could have been expected to complain that he had favored the junior co-wife over their mother. In effect, I had inadvertently done what an out-marrying daughter does, in bringing in an animal that one of my (fictive) "full brothers" would expect to use in his marriage.

Even though the animal was a female, and would probably, when mature, be milked mainly by a woman or women, no one mentioned to me any possibility that the animal might go to any of the daughters who had married out of, or daughters-in-law who had married into, the homestead. A daughter may, on occasion, receive a head of cattle (seldom more than one), smaller animals, or money, particularly if their fathers feel they have been helpful in some special way and deserve special reward. But their brothers are likely to complain when they receive large animals, particularly if these women have already married out of the paternal homestead. Rightly or not, the brothers can say the sisters have their fair share in any wealth to which these women have access in their conjugal homes. And although money is not subject to the same rules and expectations as land or cattle, it is no less contestable. When a daughter stands to receive an inheritance that competes with a son's, one is likely to hear invoked the Luo proverb *Ogwang' thurgi bor:* the wildcat's home is far away.[3] But not all women love the idea.

In deciding the distribution of property, a wuon dala is expected to try hard not to take anything *away* from anyone. From an emotional standpoint (as anyone who has cared for a child knows), it is harder to have had and to lose than never to have had at all. In these ways the culture and psychology of inheritance and devolution may depart from strict arithmetical-economic logic.

The ownership of inherited cattle is complex in itself, more complicated than mere individual property. A woman may deem herself the owner of animals inherited, but male informants insist she may not sell them, and her husband or (if she is unmarried) her father may. Luo men, and many women too, speak of the wuon dala, father or master of homestead, as master of the entire herd, even though animals are claimed by individual members—and say that only he can decide which animals enter or leave the herd. A strict taboo prohibits a young man from removing cattle without a father's permission; even if they are animals he bought himself, this is deemed a usurpation likely to bring a person or family the mysterious and deadly chira.[4]

First sons, like first wives, seem more often than not to receive bigger shares of livestock—cattle, sheep, and goats—than their followers. The same is true of land. Not only this, but fathers are more likely to consult first sons than later ones about their plans for inheritance to begin with. According to Emin Ochieng' Opere, a father who likes or trusts a younger wife more than a senior one, or a younger son more than the first one, would or should consult the favored one in secret, so as not to be seen upsetting the expected order of seniority. Some Luo elders also expect fathers to favor the sons felt most able or diligent, in the hope that they may help their less provident brothers, and may even direct them to do so.

Giving cows to married sons can also mean giving them to their wives. A man conventionally specifies his sons' wives by name in the lemo, even though it is considered (by males at least) that the sons, their husbands, are receiving them as senior claimants. The practice keeps more control in the senior generation, sparing inexperienced younger men the hazards of sparking enmity between jealous co-wives. Whatever the case, women most often end up milking the cows, sometimes with the help of boys. And the men end up plowing with the bulls or oxen, again often with boys' help as whip drivers.

As land has become scarcer, demanding more attention to its relative allocation, Luo people seem to have followed the same inheritance pattern as with cattle: equal in theory, but skewed in favor of first wives and first sons (and sons' daughters) in practice.

But cash seems not to have followed those patterns as closely. Men in rural Luo country often die (just as they live) with little cash in possession, and at least by the early 1990s, there seemed to be no hard and fast rules about its disposition. Emin Ochieng' Opere considered inherited money most likely to be deemed part of a man's *mondo*, his personal holdings (including land, granaries, and stored food) that form a kind of residual inheritance, and therefore to go to the last wife: the one likely to be least favored in land or livestock and most vulnerable as a result. A monogamist's mondo is, as noted, expected to go to the youngest son. But I have had the sense that money may be a bit more discretionary than other kinds of property, and that a man may allocate it to whomever he deems most needy, deserving, or closely associated with his own money-earning ventures.

Clearly a man deciding about divvying his property, and taking into account the likely imbalances in the needs of his co-wives or children, must consider how well they get along. My elder male informants said as much. Interestingly, though, while co-wives (*nyieke*) and jealousy (*nyiego*) are such closely associated ideas in Luo parlance—the one term coming from the other—the rivalry of male siblings or especially half-siblings, which can get just as fierce over property, is linguistically unmarked.[5] The attribution of negative qualities to women in particular is a familiar theme in Luo country. It fits with origin stories like the one about the self-activated hoe and the woman named Mikayi who, by greedily trying to make it work harder, spoiled the charm for humans forever.

Nor need a man take sole responsibility for deciding how his wealth will be divided. Luo in rural areas, and I think urban too, perceive magic and witchcraft as influencing these decisions; and a co-wife who feels hard done by is likely to suspect another co-wife (or one half-brother, another) of having used or recruited love magic or some other secret influence. Even a favored co-wife who has done nothing of the sort has incentive to lighten up her demands or even disclaim some of her share of inheritance to avoid damaging accusation or quarrel. Magical belief/knowledge may subtly serve fairness.

Many intangible things, as well as tangible ones, can also be specified for directed inheritance or succession at a pogo lemo ceremony. A man may hand down instructions about who is to become the senior manager of the homestead and its animals, or who will take on any special ritual-spiritual powers he is known to possess. He may pass on here, but usually has passed on otherwise, special knowledge of particular lineage traditions or taboos. He sometimes specifies who he expects to do which tasks in preparing for his funeral itself.

Women have not, so far, commonly held their own pogo lemo ceremonies. But they do have claims in animals and land, claims of use if not ownership; and things like these needing steady care must usually devolve to their kin before death. Most of women's human-made personal property, on the other hand, gets inherited after their deaths.

It happens at a ceremony (held for a man or a woman) marking the end of the funeral process. This is called *keyo nyinyo*, scattering iron, also lately known by a term borrowed from Swahili, *sikukuu* (holiday in the original sense).[6] Clothes and jewelry; beds, mattresses, and blankets; and pots, gourds, and other crockery and tableware all seek heirs. Things like these get piled atop the grave, or in another open place, on the day of keyo nyinyo, and after being perhaps distributed by a co-wife or other kinswoman, taken over by her co-wives, sons' wives, or daughters. Homestead members may try, but usually fail, to prevent an out-married daughter from carrying off a share to her married home. Procedures seem to vary from family to family.

The case of one young man who died intestate in Sakwa illustrates how different kinds of property may go to different kin with overlapping rights. The man lived in a sugar-growing area and at the time of his death had a field of cane grown on contract with the local sugar company, almost ripe for harvest. His widow received the cash proceeds from the cane when harvested, even though the land had already been distributed to the dead man's agnates. She also received his cattle. The man who "inherited" the widow did not, her in-laws insisted, have any rights over the disposition of those cattle left to her by her dead husband. Her own rights in these were not exclusive either, however, as brothers of the dead man also claimed their interest in them and required that if she sold any, she consult them and have one of them do the selling on her behalf. Ostensibly because of a family tradition, clansmen outside the dead man's circle of brothers and their wives were not invited to partake of the keyo nyinyo distribution. The brothers divided the dead man's clothes among themselves. One took his shirt, another his coat, and two others, a pair of trousers apiece. In all of this, the dead man's unmarried daughters living at home received nothing for themselves. This was not unfair, one of his brothers told me, since they could expect access to resources in their eventual married homes. (When they married, however, bridewealth did come in from those husbands' homes, so they found less there than there might have been.) Their sex was working against them.

Certain symbolic objects Luo people identify with ancestors or with Luo ethnic identity, and these receive special treatment in inheritance. Spears and leather shields (*kuot* if of buffalo hide, *okwomba* if of lighter cow- or bullhide) Luo associate with virility, leadership, and historic battles. Four- or three-legged figwood stools are associated with ancestors. Fly whisks and beaded hats (*ogut tigo*) are both symbols of male authority. Funeral cloaks and head-dresses are often spectacularly decorated with furry skins, ostrich feathers, hippo and warthog tusks, wildcat or python skins, and so on; and the very wildness of the animals, implying their killers' bravery, helps make them special.[7] The objects' archaic nature adds to their allure.

Nearly all these things are claimed and used by elder men and may be hedged about with restrictions. (A stool-holder's son, for instance, may not sit on it for fear of incurring chira.) They may be purchased from a crafts-man—not, customarily, by the user himself, but by a father's brother—but then they *become* things not ordinarily bought or sold. In the case of special women's garments—for instance, *owalo,* a woven cloth worn around the waist for dancing—the prospective wearer would have a brother-in-law pro-cure it. Thus a man's *pon,* or fashion piece, marked an agnatic tie, and a woman's an affinal one. The mere fact of having come to the user through a kinsperson seems to be part of what makes these properties sacred to the ex-tent they are—that is, places them outside the sphere of short-term or com-mercial exchange.

In the past, I was told, an aging man in his lemo assigned his spear or shield to a particular heir who might be outside his lineage, but this inheri-tance was unlike other inheritance because it was understood to be only for life. When the recipient eventually divided up his own property, or died, it was expected to be returned to some member of the family or lineage (*an-yuola*) that it came from: one could call it a "life spearage." According to Emin Ochieng' Opere of Sakwa and Kanyamkago in 1991 (pers. comm.), this custom was falling into abeyance; recipients were keeping the objects to pass along to their own patrilineal heirs. Such changes are hard to trace, though, particularly since some items like spears or shields that seemed archaic were suddenly pressed into military use again in western Kenya's "ethnic cleans-ing" campaigns of the late 1980s and early 1990s.

Widow "Purification" and Its Challengers

And now a topic as delicate as any. Luo have long carried out, as noted in the last chapter, a practice whereby a widow enters, voluntarily or not, into a domestic relationship, *ter,* with a brother or other kinsman of her dead husband. In a sense, a widow is considered heritable property, but in other senses not. It sometimes happens that a man giving his lemo designates an "heir" to take over his widow after his death as her guardian mate (*jater*). But three elders from Kanyamkago insisted that it is a very bad idea to single out an heir for her, and they all agreed why not.[8] Doing so places a kind of conditional curse on him and her such that if the new couple cannot get along together, no one allows the woman to leave the union. Letting her go would invite malice by the spirit of the dead man whose will was being flouted. The haunting might visit ill effects not just on her (the most likely one) or him but on any close kin or homestead members.

Widows, in a traditional Luo understanding, must be "purified," cleansed of death's contagion, by intercourse. Where this is done by an outsider residing in the oganda, called a *jakowiny* (pl. *jokowiny*)—someone morally suspect by definition—it is he who takes the risk of spiritual pollution, making her safe for an agnate to marry. This is deemed dangerous for the man who does it, as the spirit (*tipo*) of the dead, the last man to have had intercourse with her, is likely to be jealous. So the agnates may entice a nonclansman to do it. Between the husband's death and this time the widow must be sexually avoided, but once this dangerous act is done, agnates are more likely to compete for the privilege of "inheriting" her (v.i. *lago*, n. *ter*). I was told that if a widow has a baby who dies, the two deaths compounded make the first penetration of the widow doubly dangerous, and someone deemed possessed (that is, crazy), or someone who demands a large reward, needs somehow to be induced to do the deed. Some widows who feel empowered by the feminist movement have lately claimed their own right to decide whether to be inherited, and if so, by whom—much to the chagrin of some male traditionalists.

One husband and wife pair from Uyoma, interviewed in 1991, voiced their opinions about women's choosing to stick with jokowiny in preference to male agnates of their deceased husbands. Both disapproved, but for different reasons. The husband worried that a jakowiny was endangering himself and others by marrying into a lineage and clan whose particular taboos (for instance, in sequences of sex at planting time) he did not know and thus risking incurring chira by blunder. He also worried that the woman's erstwhile affines would sick their unwanted spirits, *tipo*, on him just to get rid of them. Both the husband and wife expressed concern about a woman's and lineage's welcoming in their midst someone who had left his own clan and community for reasons unknown: that he might be an outcast murderer or thief, or someone possessed by a spirit or afflicted with an incurable disease. The wife's feelings were more mixed. She allowed that if her husband were to take on a widow from another clan as a jakowiny, she would resent it bitterly, because it could damage her own reputation (and perhaps because of jealousy she did not mention). If on the other hand, she said, her husband died, and she did not get along with his agnatic kin, she would certainly welcome a jakowiny to "inherit" her instead.

As of the early 1990s, most widows still acquiesced to, or actively chose, being inherited, not least for protection. But public health workers, concerned about HIV/AIDS, were crying out against widow purification rituals and widow inheritance altogether. By late March 2001, the Nyanza provincial commissioner, Peter Raburu, issued a public threat of arrest for anyone who forced such a rite.

Practices like widow inheritance and bridewealth die hard. Attempting to regulate them by law pits the authority of the state against the authority of the elders and ancestors—unless of course these last can be persuaded to change their minds. And lately the state has seemed to many in the western hinterland distant, unsympathetic, and weak. But this time, for once, the state has had church missions on its side.

Christian missionaries in Luo country, and some Luo Christians influenced by their example, have expressed to other Luo their discomfort with— and sometimes frontally assaulted—aspects of inheritance customs and perceptions that seem to them to represent ancestor worship or clannishness. In particular, the idea that disregarding the will of a dead person brings a curse upon one or one's kin has been repugnant to some Christian sensibilities, since it suggests that there are living spirits of the dead who may compete with Jesus Christ, a holy spirit, a supreme deity, or ecclesiastically designated saints. The Luo concept of lemo, if translated as prayer rather than merely as declaration of a will, is likely to strike some Christians as prayer misdirected. Apportioning out one's worldly goods to only a limited number of near and dear also appears to some church people inconsistent with a Christian ideal of universal brotherhood and sharing. Christian clergymen who want to be called "father" do not welcome competition from other fathers or grandfathers, particularly elderly or dead ones who, in the gerontocratic order, are entitled to more authority than they. Particularly objectionable to those who disagree with indigenous ways are Luo patterns of "widow inheritance," which seem to some to deny a woman's free individual will and to foster acts such as "purification" by intercourse with a stranger (jakowiny), which, when followed by a union with another man, raises in Christian minds issues of profligacy or adultery. (Indeed, this last is how some translate ter-ruok, the practice or condition of widow inheritance.) Christian dogma here mixes with issues of women's rights and public health. These are political as well as spiritual issues. Feelings about them run to extremes.

It would be too simplistic, however, to say that these issues neatly divide Luo from Christians, or modernists from traditionalists. Some Luo who identify themselves as Christians (for example, members of the Children of God Regeneration Church, in Kanyamkago in the early 1990s) felt strongly opposed to most other Christians' disparaging teachings about local practices deemed traditional. Christian denominations in Luo country are countless — always dividing and subdividing like branches of a tree, just as lineages do — and practically any custom deemed objectionable by some is tolerated or even embraced by another. And yet hard feelings remained. For support against Christian anti-indigenous campaigns against funerary and inheritance tradi-tions, some Luo looked to state-sponsored organs like the Kenya Broadcast-ing Company on the radio, or the dancing troupe Bomas of Kenya. Among people usually isolated at the margins of a sometimes hostile state and jeal-ous of their cultural autonomy, these may be signs of desperation, but they also show ingenuity.

A Delicate Affair

All in all, inheritance is a tricky thing. Luo know it to be a game of navigating conflicting principles. You can be evenhanded or you can re-spect seniority. You can get a widow inherited and placate her future spirit before she dies, or you can avoid the wrath of the dead man's own cuckolded spirit who may object. You can partake of a sacred indigenous tradition or satisfy Christian missionaries and maybe avoid some sort of damnation. So often, in decisions about inheritance, do humans have to steer between some Scylla and Charybdis that it becomes hard to imagine a perfect inheritance scenario — or anything but a brittle set of compromises, a trusting in forgive-ness — in what Luo call *werruok*.

Luo inheritance is about not just goods and services but also about bads and disservices. Luo inherit debts and old scores. But it is also about people themselves. What land you inherit, and from whom, helps define who you are, and you may be inherited yourself. Who benefits, if anyone, may not be obvious. What looks on the surface like land passing from father to son is often, in practice, land passing from mother-in-law to daughter-in-law. While some outsiders perceive "widow inheritance" as a kind of forced enslavement calling for liberation, more than a few women in equatorial Africa continue to accept it, even demand it, leaving open to speculation whether it is they, their peers, or their ancestors who are doing the talking.

The strivings in inheritance decisions may seem ingenious to some, futile or diabolical to others. Invoking metaphorical adages like *ogwang' thurgi bor,* the wildcat's home is far away, to justify sex bias may help salve a troubled male conscience or temporarily quell dissent, but it may not please the women alluded to. Having a stranger "purify" the widow before inheriting her may not please the public health officials, but it is what some local adepts, concerned as much with spiritual as with biological hygiene, prescribe. Inventing a new church with a new mix of African and Christian morals has often seemed to Luo the easiest way to cope with the contradictions while staking a claim of one's own, as a breakaway lineage stakes a claim to becoming a future clan or nation. These are all chancy propositions, but what they do not solve, some hope, they may transcend.

In the lake basin, again as elsewhere, a man's death and the distribution of his property are the occasion that is more likely than any to sharpen family animosities and lead to lineage fission. Some would say there is a deep psychological element at work here: bereaved persons who may be, at one level, grasping for the person departed, end up grasping for the land or property as a surrogate, and investing more emotion in it than they would if it came from another source.[9] Be that as it may, land, cattle, and ceremonial paraphernalia continue to carry more sentimental value than cash does in Luo country. Identities are fused with them. When, over land or cattle, kin shed blood of kin, few Luo are likely to be surprised. When interests of clansmen come into conflict with interests of outsiders, as they seem to be doing increasingly in cases of widow inheritance, few may be surprised at an otherwise inexplicable disease or accident that ensues. And of these, the Luo country has lately seen plenty.

A reminder in Luo inheritance is that there are some obligations that can be discharged only by passing along of what has been received. A hidden key word in this is "of." For it is not always possible to pass along the very thing given us, in the very measure. If the family has grown, the land may have shrunk. If one had to move in a hurry during an ethnic pogrom, what is left may be the pouch of gold dust. If the gold got pawned or stolen, what remains may be a word of wisdom. Inheritance is plastic; each generation can change its form. But to play the game, to convey an entrustment from forebears to heirs, is to bind past to future, securing one's own place in a sequence and a process. It is to enter a flow.

Partaking of that fluid tradition, humans sometimes feel obliged to toss something back from whence it came: to feed an ancestor, to chip in for the alma mater, to cite a source. How Luo and other eastern Africans make sacrificial offerings, and why, is the subject of the following pages.

CHAPTER 9

Blood, Fire, and Word
Luo, Christian, and Luo-Christian Sacrifice

Religious faith and superstition are quite different. One of them results from fear *and is a sort of false science. The other is a trusting.*
—LUDWIG WITTGENSTEIN

Perhaps no human action commands more fascination, or is surrounded by more mystery, than sacrifice. Few things occasion such high emotion, so much suspicion and misgiving. What one considers the highway to divinity, and a lifeline to what is real or eternal, another considers a sullying sacrilege or a proof of superstition. What one deems a calculated gamble, an investment, or insurance, another sees as an escape from all worldly concerns. What one thinks of as a chance to eat some meat for a change, another perceives as an abuse of animal, maybe human, rights. But real people often express mixed feelings where sacrifice is concerned. The difficulty of reconciling them helps keep the topic close to the center of human concerns, yet it frightens some even to think about it.

If humans can place trust in spirits or gods, they can also entrust things to them; and when this entrustment involves things with a material dimension, it deserves consideration as part of the fiduciary culture that is the topic of this work. It was suggested earlier that in agrarian communities in tropical East Africa, death is never far away, but humans keep the dead and nonliving in one way or another more alive, and more active participants in daily life, than they might seem to some in other parts of the world. If humans make entrustments among themselves, are they doing something comparable when they make sacrifices or offerings to spirits or to divinity? What becomes of the sense of reciprocal obligation, and what do people think and do if sacrifice and prayer are not answered as hoped? Where people in the past have counted upon the sacrifice of animals to communicate with spiritual beings

or forces, how easily do they adjust to new faiths whose adherents profess another form of sacrifice or offering as a sole, ultimate, or at least superior way
and try to expunge what the people have long regarded as their most fitting
and perhaps most reliable recourse? After briefly circumscribing this complex topic of sacrifice, or what is so called, I describe some older and newer
forms, and what they have to do with each other—quite a lot, it turns out,
even if we just confine ourselves mainly to those aspects bearing on entrustment and obligation.

Sacrifice: A Nilotic Preoccupation?

Anglophones who speak of sacrifice (from the Latin *sacer* and *facere*
—to make sacred) commonly consider it a religious act. Evans-Pritchard
more than once called sacrifice the "most fundamental rite" of "religion,"
and other noted anthropologists and sociologists have agreed.[1] But sacrifice as
we know it today is by no means *just* religion.[2] Sacrifice is a guarded privilege
and a way to seek or exercise power and privilege (and therefore political), a
dramatic performance (and therefore aesthetic), a way of raising hope, salv-
ing conscience, or perhaps projecting or transferring aspects of the self (and
therefore psychological). It is often, too, a way of joining people together, and
in doing so potentially marking them off from others (hence sociological).
If it channels cruel or aggressive impulses into an object who cannot strike
back at its enactors or their progeny (a domestic animal held down or tied
up) and then provides some nutrition, it may be called biologically adaptive.
Combining so many elements, approachable from all so many angles, sacrifice comes as close as anything to what Marcel Mauss called a "total social
phenomenon"—or just a total phenomenon. As such, sacrifice attracts the
attention of disciplines that work (so to speak) at cross-purposes. The discussion that follows does not purport to make a statement about ultimate
truth, nor to cast judgments, but only to outline briefly some features of a
way of thinking found in the Luo country, Kenya, and elsewhere: one that has
proved flexible, adaptable, and persistent. Part of what makes these ideas endure and recur is that the ideas are both symbolic and economic, and in some
ways also political and aesthetic. These are aspects of life that many live their
lives, and conduct their sacrifices, without breaking apart in the first place.

People of the middle and upper Nilotic region are famous for sacrifice.
They first became known to the world through the Shilluk institution of sacrifice (and self-sacrifice) of an aging or weakening priest-king, as reported

by German missionaries P. Banholzer and J. K. Giffen from about 1905 and publicized through James Frazer's *The Golden Bough*. A roughly comparable pattern reported among Dinka (Lienhardt 1961), involving a semiofficial spiritual leader called a spear master, and the Nuer sacrifices of cattle and other animals, made famous by Edward Evans-Pritchard (1956) and restudied and reanalyzed by many others, has made the stuff of an anthropological tradition that has made upper Nilotic culture almost synonymous with blood sacrifice.[3] Is this tradition some kind of intellectual "mad cow disease" that comes from too much consuming of our own kind, or do Nilotic people really do little but go around thrusting sharp-pointed objects into hapless hoofed creatures?

Luo in Kenya, who have become less heavily dependent on cattle keeping than some of their kinfolk to the north in Sudan have remained, have never struck me as a people obsessed with sacrificing animals, let alone humans — at least not in their daily lives. A busy, engaged person can go for weeks, even months, in the Luo country without seeing a single instance of sacrificial slaughter. (Mostly one just sees a butcher here or there, slaughtering animals for sale on an open-air slab somewhere near a marketplace.) And yet it does go on. The most likely public occasions to see it done ceremonially to cattle are at funerals — where, however, not all animal killing could neatly be classed as sacrificial, since some is done there without special ceremony just to feed the many attendants. Smaller animals are killed in sacrificial offerings more privately. Luo people make offerings of animals, beer, and other things, on their own behalf or on behalf of others present or absent. At the dawn of British colonial rule, administrator C. W. Hobley described Luo goat and sheep sacrifices to mend a broken taboo or to appease or placate spirits during illness. He noted feasting and the throwing of small bits of meat, and the blowing of beer or millet porridge, in four directions to appease or placate spirits (1902, 31–2; 1903, 343; see also Northcote 1907). John Roscoe (1915, 286) and others also briefly remarked on goat or sheep sacrifices at graves by the fathers of persons sick. Some of these early descriptions read much like standard descriptions and prescriptions given to me by Luo themselves toward the century's end. One young secondary school leaver, for instance, from a family of modest means in Komenya Location, Kagan, in 1981 described what he thought should be done before moving away from a home where there has been illness or death, so as not to be followed by malevolent ancestral spirits causing the harm. A black, white, or black and red hen should be killed and roasted whole, he said, or beer brewed.[4] Some of the meat or beer should be thrown onto the graves of ancestors for their spirits, and some toward the sun for Nyasaye, to eat.

Christian missionaries in western Kenya, as elsewhere in Africa, have tried hard for a century to stamp out animal sacrifice to local ancestors or spirits. They have sought to draw attention instead to their own biblical tradition and to what they considered "the Holy Land." Their scriptures include many references not only to sacrifices of four-legged animals but also to offerings of humans, notably Abraham's offering of his son Isaac, spared once the willingness had been shown. References to Jesus Christ as a "Lamb of God," and the enactment, in some churches, of the Eucharistic sacrament in observance of the crucifixion, are the things to which many Christian missionaries and their converts and followers have turned attention. And yet more than a few Luo people and others in and around western Kenya have continued to offer their *own* animals, home-brewed beer, and other valued things, as is said to have been practiced since earliest known times—to ancestral spirits, other local spirits, and divinity. Most of it happens beyond the gaze or ken of Christians expected to disapprove. But that does not mean no Judeo-Christian elements have entered into the thoughts or practices involved.

So let us look more broadly at what kinds of Luo practices and understandings might be classed as sacrificial in the first place.

Forms and Idioms

What Anglophones call sacrifice is known in Luo by several terms with overlapping uses but varied shades of meaning: *liswa* (n.), *misango* (n.), *dolo* (n., v.t.), and *msiro* (n.). Of the first two terms, *liswa* is often used to refer to an offering by elders to the earth (in a communal sacrifice) or to spirits deemed to be demanding action by bringing trouble. *Misango* most commonly refers to a ritual act, sometimes a killing, performed by an invited *jabilo* (specialist in "medicines") to cleanse a household (for instance) or its inhabitants of a curse or disease, or to reverse some other unwanted condition; but the term is also sometimes used to refer to an act in a rite of passage such as a wedding or funeral. Since liswa or misango can affect a place as well as the people in it, some Luo have preferred it be performed by weg lowo, that is, the prior occupants or landholders (that is, of an oganda) or their descendants, rather than by jodak, land clients or borrowers. Or some, at least, have had a vested interest in reserving the privilege to them. *Dolo*, like misango, can refer to a sacrifice or offering in one of the smaller and more private contexts, and it usually implies a smaller animal (for instance, a quail or small chicken) rather than a large ruminant. Or dolo can refer to a poured libation. Dolo and msiro, performable by either weg lowo or jodak for their own homesteads, are meant to bring the order of things done (in time, in space, or both) within the proper or "natural" order. When a prohibition known as *kwer* (roughly translatable as taboo) is broken—for instance, by a man's sleeping with his wives in the wrong sequence at planting or harvesting time—dolo or msiro may be called for.[5]

Blood sacrifice, translated in one way or another, has long been a standard part of birth and naming, wedding, funeral, and other transition rituals, and common practice in healing and other vital supplications.[6] Generalized Ocholla-Ayayo as recently as 1976, "For all the Luo [including Sudanese and Ugandan Lwoo] tribes, i.e., the Northern, Central, or Southern Luo, every bull or ox is ultimately destined for sacrifice or an honorific ceremony," the bigger and more public and collective events for Nyasaye, the supreme divinity, and the smaller, private ones more often for ancestral spirits (1976, 170). This was an exaggeration (since some cattle were sold to butchers or urban-based traders), but it does reflect a Luo and broader lake-river Nilotic ideal. Not just cattle but sheep, goats, and chickens are common objects of sacrifice. Where a small animal does not bring a desired result, a larger one is sometimes slaughtered, as if by negotiation in a marketplace.

In some respects Luo "custom" for sacrifice and other slaughter, in thought or deed, active or dormant, can be remarkably complex. Witness, for instance, the finely detailed "cow map" with scores of named animal body parts due to particular categories of persons in ritual meat distribution—for example, *chok-agoko* (sternum), "reserved for the senior boys, eating it in a group" or ones remembered as so designated in the past, as in *chok odiere* (humerus), which "used to go to the house of the senior wife or the one next to her in rank."[7] Moreover, sacrifices for particular purposes ideally require animals of particular species, particular sex and sexual status (castrated or not), and particular color and color combinations (white ox, black ram, black and red hen, and so on), though Luo people do make substitutions to get by with animals and substances available.[8] All this adds up to a lot to remember, and nearly everyone in Luo country can, if pressed for details, tell you of some old *dani*, grandmother, in the next valley who knows the ropes better. The specificity of requirements for a "successful" supplicatory sacrifice means that in practice, enactors must surely bend the "rules" and make compromises or substitutions often. All this no doubt helps perpetuate the understanding of sacrifice as efficacious by giving potential reasons why it might have failed in particular cases where accompanying prayers seem not to have been answered and nothing improved.

To whom, or what, do Luo pray and sacrifice? The short answer is: both the one and the many. Since the early twentieth century when systematic records began to be kept, Luo have spoken of "the one" as both a supreme, all-embracing creator divinity, god, or power, lately most often referred to as Nyasaye but also known as Were (provider, merciful) or sometimes as Nyaka-laga (the omnipresent), Jakwath (guardian), or sometimes Chieng', the term for sun. There has been some debate, including by Luo scholars themselves, on whether a supreme deity is an indigenous Luo concept or was introduced by Christians or others in European times. *Juok,* a second key concept in Luo faith and cosmology, is sometimes also used to refer to a personified spiritual entity that ramifies into the many "lower" spirits (*juogi*) or powers; but it is sometimes used to mean an impersonal one (comparable to *mana*). The spiritual powers, in turn, are sometimes divided by scholars into the tutelary (that is, associated with or protecting specific human groups or individuals) and independent or free-floating ones. Some of the tutelary and free spirits (if so they may be called) are associated with specific features of the inhabited material world—for instance, unusual rock outcroppings, springs, large snakes, and so on. Ancestral spirits may be seen as greedy, jealous, or vindictive, and yet they are also deemed watchful to ensure harmonious social relations among their living descendants.[9]

Animals are slaughtered sometimes not as offerings to particular spiritual beings but instead in a more direct way to right wrongs, to correct breaches of taboo (such as incest or father-beating), or to seal pacts—or to dramatize these changes. On some of these latter occasions—for instance, in peacemaking ceremonies noted in the past—chyme (stomach contents) of a slaughtered animal has been spread on the bodies of participants as if to emphasize their unity or to externalize what has been kept within and thus invert an unwanted condition. This is a custom that Luo and Bantu-speaking neighbors have shared.[10]

As all this implies, the acts that get called sacrifices, in Luo country or elsewhere, vary enormously in form, intent, and interpretation.[11] People sacrifice or offer live animals, beer (*kong'o, aput*), or other things by graves, at the gates of homesteads, or under large trees outside. But they can do it almost anywhere they consider spirits to reside. They may do it at predictable life-cycle crises like wedding or funeral celebrations, or anytime an illness or crop failure seems to pose a threat. Sacrifices can be considered voluntary or obligatory. They are performed by solitary persons or by groups; with animal or vegetable objects. When addressed to spirits, sacrifices may be purported to bring these closer, but more often are to keep them and their dangers away. Although Luo seem to sacrifice as often as not to protect themselves, they can do it to procure, thank, restore, atone, expiate, transcend, or a mix of these. Often there seems to be a general idea of restoring disrupted relations between living and the dead—for instance, when people are quarreling and the living suspect the dead of feeling neglected and causing dissent. Practical purposes of sacrifice can include (to name only a few) sealing a marriage or brotherhood; procuring rain, children, or other good things; warding off illness or other danger; ending a plague of locusts; protecting a fishing boat or a pickup van and those who ride in it. Mood during a sacrifice may be joyful, somber, or anxious (and it may lighten up dramatically upon the completion). Where an animal is involved, techniques of killing can vary from spearing to beating to suffocating. Sometimes blood is dripped into the ground, sometimes flesh sent up in smoke. Sometimes participants consume the material; at other times, when a calamity as big as a plague or drought is at issue, they do not. The economic costs can vary from nearly nothing (as in a breathy burst of mouth water or a gourd of grain beer) to at least several head of cattle. The benefits are calculable only in the eye of the beholder.

No single criterion for "sacrifice" adequately captures all the acts described above. A substitution of one body for another, a giving up of something for something greater, an act of violence on a helpless victim, a killing and offering of a life, a transferring of its perceived power, a shared ceremony and feast . . . any of these may be part of "sacrifice," but not all need be present.[12] A sacrifice, in short, is something that has enough salient features in common with other things called sacrifice.

Sacrifice is easy to misconstrue. Skeptics who doubt the sincerity of sacrificial acts by noting that the humans end up eating the meat of a slaughtered animal, or most of it, sometimes neglect that it may be the life, strength, or breath of the animal, rather than the meat, that is considered the thing being offered. Churchpersons who call sacrifice idolatry or pagan worship sometimes misconstrue sacrifice as an act of worship of an animal rather than the use of an animal as a channel of communication with a spirit or divinity. A half-dead animal seems to serve just this latter symbolic function for some: suspended momentarily between the living and nonliving, it serves as a kind of portal through which to contact the other side.

Whatever form sacrifice takes, when it is directed at a divine or spiritual entity deemed conscious or willful, it seems usually to be performed with a prayer that a benefit, or a combination of benefits, will be granted in return. It can therefore be called an entrustment. It is not perhaps understood to compel a favorable response so much as to ingratiate: not to obligate but to oblige. Sacrifice and prayer can serve simultaneously the double purpose of thanking for past favors and soliciting future ones.[13]

Sacrifice can also help control the experience of that which is not in people's own control: to feel as though they have at least done what they could. Since only some persons, typically genealogically senior males, are expected to sacrifice on behalf of their families or lineages, their doing so can serve to announce, rehearse, and perhaps reinforce a power differential between the genders and a gerontocratic order. Sacrifice, like other ritual acts conducted with prescribed sequences and formulae, can give the living a feeling of being bound with the dead and the unborn. In doing so it underlines their membership in a particular community.[14]

Details of procedure are always important, and having complexity inherent in the oral or written "rules" and "secondary elaborations" like sex or bathing taboos allows for easy explanation where sacrifice seems to fail to achieve its ostensible ends. These elaborations may not be so secondary: without them, a structure of understandings (belief/knowledge), and of action that comes with it, would be easier for skeptics to discredit. The careful sequencing of ritual action itself helps define the bounds of a community of practitioners; to conform to an expected sequence is to signal willingness to conform and fit in. A set sequence also reminds the living of the dead and unborn who are supposed to have done and will do things the same way.

Now about human sacrifice. This is a delicate topic that can be mentioned only briefly here but would be easy to sensationalize. Some Luo, including college-educated ones, have told me of unnamed others (including, in the past, *jobilo*, or magician-diviners of *ogendni*, clan federations) who in their view have used their powers of witchcraft or sorcery to conduct the sacrifice (*misango*) of a child, out-married woman, or other victim in the expectation of getting good herd health, rain, or harvest in return, through the agency of spirits and through spirits, divinity. In the past it was sometimes done upon moving to a new area. In our times, I was told, it is done upon planting or harvesting, both to thank for a good harvest and to supplicate for another. The magical sacrifice of humans is something I have heard no one admit to attempting: invariably the finger gets pointed at some other place than where the speaker happens to live or come from. But the knowledge, belief, or sup-position that such magical human sacrifices and tradeoffs occur is no less alive for it. And comparable understandings have been found among at least one other neighboring people across the Nilotic-Bantu linguistic divide.[15]

Underlying the sacrificial discourse and practice in Luo country, as else-where, are two important assumptions. One is that there can be *no gain with-out pain.* The second is that *one being can symbolically substitute for another.* Both assumptions, it may be recalled, also underlie the understanding of "bitter money": money that, because it has not been earned by honest labor, can bring only harm to its owner or kin. In the logic of sacrificial ritual, cer-tain tenets of ordinary reasoning get suspended. The symbolic substitution of one thing for another does not depend on one-to-one correspondence. One sacrificial victim can shed blood for many. Moreover, in sacrificial logic, one can suffer, be offered, or self-offer for others in past or future.

That sacrifice can establish a kind of contract (or the perception or feeling of one) between the one sacrificing and the one sacrificed to has been an im-pression held by sociologists and anthropologists since set down by Hubert and Mauss more than a century ago (1981, 100), who generalized, "Fun-damentally, there is perhaps no sacrifice that does not have some contrac-tual element: The two parties present their services and each gets his due." Among Nilotes in particular, the idiom used to discuss sacrifice has some-times been translated with concepts including purchase, exchange bargain, ransom, redemption, and indemnification (Evans-Pritchard 1956, 221–25, 276; cf. Sperber 1987, 14–18). Writes Fortes, of cases as widely separated as the Tallensi of northern Ghana, Nuer in southern Sudan, and Assamese in India, "there is generally an implication of sacrifice being obligatory *as if in payment of a debt,* neglect of which could be disastrous," rather than in the spirit of spontaneous love (Fortes 1980, xv, my emphasis).

And yet the use of such idiom does not necessarily imply the same kinds of sanctions that a contract among living humans can, since humans cannot punish spirits in the same ways that spirits can punish humans. Nor does the idiom of contract adequately describe the range of emotions — of identifica-tion, solidarity, fear, hope, release, or transcendence — that parties to sacrifi-cial ritual may experience before, during, or after a ritual sacrificial act.

There is a further point about sacrificial entrustment and obligation that is often overlooked. A sacrifice can create a debt not just between the one sacrificing and the one sacrificed to, but also between them and some third party. That is, A's sacrifice of object B to "supernatural" agent C also in-debts some perhaps unspecified D. When Christian Luo or other Christians, for instance, contend that "the Savior died for our sins," they evoke a feeling of indebtedness among the living to whom they speak: an indebtedness to the savior, if not also to divinity or the sacrificer(s) (who, by some interpretations, is in this case the selfsame savior, self-sacrificed). The obligation ramifies, and living observers sometimes construe their common indebtedness as a bond between them. Sacrifice can thus be construed as part of social fabric.

All these considerations — the symbolic substitution, the vicarious suffering, and the multiplication in the extended covenant or contract with a sense of debt or obligation involved — come vividly to bear in the faiths in which Luo have participated since the arrival of Christian missionaries. These include both faiths that are anti-Christian or non-Christian in doctrine, but that can still incorporate Judeo-Christian elements nonetheless, and contemporary Luo Christian religion, with its rich mix of indigenous and exogenous elements. We look at these now in turn.

Sacrifice in Protest: A Nilotic-Bantu Purification Movement

In at least one dramatic episode, some Luo and neighboring people in Kenya have been tempted to give up their cattle and other livestock en masse in the hope and expectation of greater rewards to come. It took the form of a protest movement — call it religious, political, or socioeconomic.

It began within only about a decade of the start of local colonial rule — that is, since the establishment of a district headquarters at lakeside Karungu in 1902 — and the rather arbitrary ensuing designation of new chiefs. It followed more than a decade of forced labor in road and railroad building and the imposition of new taxes. It began within a decade, too, of the start of Christian evangelism with the establishment of the Seventh-Day Adventist mission at Genia in 1907 and the Catholic Mill Hill Fathers mission near Kisii in 1911.[16]

The movement took firmest root among Luo and later among (Bantu-speaking) Gusii after the Kenya African Rifles and police had conducted violent punitive expeditions (some after minor incidents of cattle rustling and a killing or two) in the latter's home area of Kisii between 1905 and 1908. In a January 1908 incident G. A. S. Northcote, a moderate noted earlier who by then was the South Kavirondo district commissioner newly based in Kisii, was himself wounded by a spear in the back; British military commanders used this incident as an excuse for a violent campaign of retribution. Scores, and probably well over a hundred, of spear-armed Gusii were massacred with machine-gun fire and bayoneted rifles—the British suffering few or no losses (Maxon 1989, 42–43). From Gusii were at least temporarily confiscated many thousand cattle, sheep, and goats (many though not all of which would eventually be returned to them, some in exchange for labor).

In brief, a new British presence, political reorganization, forced labor, Christian evangelization, new taxation, wholesale confiscation of valued animals, and crushing military defeat against previously unknown weaponry were among the preconditions for the movement's spread.

The protest movement arose as what some have called a cult but others would call a faith community: a movement of locally constituted groups, loosely organized, with an evidently shared vision and set of practices. It took the name of Mumbo, a giant water serpent—in some accounts, one stretching from the lake into the clouds. The idea for this figure in turn was perhaps inspired by the pythons native to the lake basin, or by the water spouts from cyclones that sometimes appear over the great lake.

One evening in 1913, one version of the legend goes, Mumbo appeared to, and swallowed ("possessed"), a Luo man from Alego named Onyango Dunde, regurgitated him, and spoke to him. The snake proclaimed, "I am the God Mumbo whose two homes are in the Sun and the Lake. I have chosen you to be my mouthpiece. Those whom I personally choose and also those who acknowledge me, will live together in plenty. Their crops will grow of themselves and there will be no need to work . . . but all unbelievers and their families and cattle will die out." "Christianity is rotten [mbovu]," and "Europeans are your enemies," Mumbo told him to tell the people of Alego in particular and Africans in general. Mumbo's words included commandments and a conditional promise to relay. "Daily sacrifice—preferably the males—of cattle, sheep, goats, and fowls shall be made to me. More especially do I prefer black bulls. Have no fear of sacrificing those and I will cause unlimited black cattle to come to you from the Lango, Maasai, Nandi and Kipsigis. When this is done, I will provide them with as many more as they want from the lake" ("Nyangweso" 1930, 13–14; Wipper 1977, 35). If they did this and followed other instructions (wearing skins and long hair, refraining from bathing) precisely, some were induced to believe, they would gain unlimited wealth in the new age soon to come, an age in which people in and around the lake basin would be rid of Europeans, foreign influences, and evil. This was a nativistic movement in at least two senses: seeking to redignify local culture and tradition, and to relocate the center of spiritual inspiration and authority closer to home.

The Mumbo movement that arose on the basis of Onyango Dunde's prophecy, later followed by others, spread among some sizeable minority, especially women and especially of Luo and a few Kuria living to the south of the Winam Gulf of the lake, and then of the Gusii up in the hills to the east. Precipitating the spread, it seems, were conditions and events of the Great War. These included the hardships that western Kenyan people underwent in forced labor for the dreaded Carrier Corps. They also included a demonstration that British presence might not be invincible. The war's tensions reached directly into western Kenya as Germans briefly invaded Kisii in 1914, causing British administrators to flee, and then as German Christian missionaries (Seventh-Day Adventists and others) were sent away or interned. During the war years from 1914 to 1918 and afterward the Mumbo faith or cult spread rapidly. "Some followers, believing Mumbo's promise of food in the millennium, stopped cultivating their gardens; a few killed all of their livestock as he had ordered"; and others stopped helping government officials build roads (Wipper 1977, 44). Colonial administrators and their agents who investigated the meeting centers that Mumboists built reported finding material evidence of sacrifices conducted there, and these sacrifices seem to have been part of what frightened them about it—they mostly deemed it perverse, irrational, and superstitious—and motivated their repressive responses. These included rounding up the prophets or priests they felt were its leaders and sending them to the coast for detainment. The Mumbo movement seems to have reached its peak in economically depressed and stressed times of the early 1920s and early 1930s. Its practices were banned by the Local Native Council but revived for brief periods around 1938 and 1947, and the 1950s, being banned again in 1954 in the period of Mau Mau emergency (Wipper 1977, 38). Available sources give scant indication how many practitioners or adherents the Mumbo movement may have had. We know British colonial authorities deemed it a threat, but since government officials often do panic and forcibly crush politico-religious movements whose evident mysticism puzzles and scares them, it is hard even to tell whether this movement's numbers and any weapons possessed were such as to pose the colonial regime any real threat at all.

The Mumbo movement incorporated elements of both local tradition (sun focus, spirit possession, special value on cattle) and the then-novel Judeo-Christian tradition (Jonah and the whale perhaps; revelation of a new earth to come, with a judgment and a threat of perdition for unbelievers). In this and in other respects the Mumbo movement is reminiscent of a broad range of other movements originating elsewhere. They take many labels—nativistic, millenarian, messianic, and apocalyptic, for instance—which sometimes, however, carry baggage of culture-specific implications from other parts of the world. These have sometimes been collectively called "revitalization" movements (a term of Anthony Wallace's), but in my view they could as well be called "redignification" movements. Typical is some sort of promise that an existing social order will be upended and "the last shall be first."[17] In recent times they have typically been found among invaded, colonized, dispossessed, disempowered, and otherwise oppressed or humiliated people; many examples come from central Africa, Melanesia, and the American West. Typically, they focus on a charismatic visionary prophet, become insular in communications, become rightly or wrongly perceived as a threat to bewildered or spooked government officials, and are brutally repressed. In a way though, for its participants, Mumbo seems to have been an attempt to restore a lost local dignity, and to make sense of the disjunctures brought about by competing worldviews, by breaking them up and melding together elements from them into a more cohesive, if utopian, narrative and vision. Sacrificial entrustment was a central part of that vision.

Related to, but distinct from, the Mumbo movement is another western Kenyan movement focusing on a "real" (living, mortal) python named Omieri (or Omweri)—again by the lake shore around the Winam Gulf. This movement has surfaced sporadically over decades, with the appearance and reappearance of one or another snake so labeled—more often than not female, sometimes with brood—deemed propitious for rain and good times. The snake, here again, has been an object of reverence, indeed worship. The movement has surfaced especially during times of drought and famine in the region—including the time of this writing—and it has been a focus of intense but hitherto peaceful controversy involving churchpeople, wildlife officials, and the interested public. The faithful observance of Omieri—thus far generally peaceful—by some western Kenyans could be called a nativistic movement, and one of revitalization or redignification. Some claim to have observed sacrificial offerings to Omieri, but not on the scale reported demanded for Mumbo, and in this respect sound information on both remains scarce.

Mumbo was an avowedly anti-Christian movement; Omweri has thus far been a non-Christian one. We turn now to an avowedly Christian one, of a sort that has grown and multiplied in a wide range of independent churches with local charismatic prophets at their base—in this one a martyr, ostensibly self-sacrificed—to see entrustment as an item of faith.

Sacrificial Covenant in an East African
Christian Church Movement

Some Luo people say there are Luo saviors, who have died sacrificially for others' sins in a manner they have come to compare or assimilate to
that of Jesus Christ. One independent Christian movement, built around the
memory of such a figure, is known by the name *Roho*, or spirit. It is discussed
here not as an instance of doctrinal truth or falsehood but as an instance of a
Luo and African Christian faith with ideas of entrustment and obligation at
its heart. In what follows, a long story is made brief.[18] It suggests, among other
things, how thoroughly indigenous and exogenous religious understandings
can mix — particularly where, as in this case, existing knowledge of the factual history is sketchy and biblical teachings are available to people to help
fill in missing details. It also suggests how little separation there can be between religion, politics, and economy in real life.

Alfayo Odongo Mango was born at an unknown date by about 1884, in
the Kager clan of Luo, who were engaged in a long border dispute and skirmishes with Wanga, a subgroup of Luhya, after a history of movements back
and forth. Son of a displaced father, and eventually evicted himself, he grew
up to be indignant at British appointment of chiefs and headmen over Luo
homes and communities — particularly the paramount chief Mumia — and at
the colonial drawing of a boundary that placed Luo in Wanga territory.[19] Like
many who become prophets, he was an insider-outsider in his society, in his
case experienced in trade, mission schooled by Anglicans at Maseno, and
patronized by the celebrated archdeacon Walter Edwin Owen (with whom he
eventually fell out as he traveled and gained his own following). In Mango's
life and the new movement, spirit possession and healing — long aspects of
Luo tradition — played a big part, and these are women's traditions no less
than men's, and indeed more. Just as Luo lineages branch apart, a new independent Kenyan church branched off an older, Anglican one. Mango continued to challenge the Wanga paramount chief's hegemony, the British support behind it, and the land patronage over Luo that came with it. He died
on January 21, 1934, at Musanda, in one of a series of Luo-Luhya skirmishes
that some call a massacre. He died speared — but reportedly without resistance — in his own burning home, and by some reports his body was found
half charred.

Around Alfayo Mango's martyrdom a church and faith crystallized, with
Musanda as one of its pilgrimage sites. Mostly women joined, and some dis-

tinguished themselves in church leadership, but men eventually took over most of the high church offices. Today adherents of *Dini ya Roho*, the religion of Roho, have their own distinct robes (white, with prominent colored crosses), hymns, and a secret "language" of encoded Luo terms. Named subdenominations have continued to branch off the stem.[20] Roughly estimated at 50,000 to 75,000, plus unregistered followers, in the mid-1990s (Hoehler-Fatton 1996, 216), Roho followers have formed churches in Kenya, Uganda, and Tanzania, with missions expanding. No longer just a Luo movement, Roho now includes Bantu speakers too.

At the core of contemporary Roho interpretations of the story is the Luo concept of *singruok*, contract or covenant, and exchange is an integral part of this concept. Mango and the eight others burned in the fire are understood to have agreed to offer themselves up in death, and in return the Holy Spirit would remain and fill their land instead of going somewhere else (Hoehler-Fatton 1996, 129–30, see also 143, 156–57, 229n; Ogot 1999, 128). Now singruok, we might add, comes from the root *singo*, which Grace Clarke translates as "Security . . . Anything lent, given or left with one for a specific purpose and a limited time" (n.d., S.10).[21] This would make not just entrustment and obligation, but actually credit and debt, seem apt terms for the pact as recounted in oral history. Singruok also invites the idea of "redemption," in both its economic and spiritual senses, recalling Christian understandings too about the sacrifice (or self-sacrifice) of Jesus Christ and the salvation of humanity (Sykes 1980, 66, 70–72). To Roho followers, Alfayo Odongo Mango is roughly "at par with Christ" (Ogot 1999, 128) and in a sense they may partake of each other. Roho followers sing of beholding the salvific fire (*Ne War Mach!*), of trusting (*geno*) Jesus, and of drinking the blood (*remo*) of the new Israel (Hoehler-Fatton 1996, 151). Sometimes they speak as if it is African or dark-skinned people who will be saved by the sacrifice, sometimes it is all people.

Some will celebrate, and others lament, the past century's proliferation of Christian traditions (and Islamic ones too) in Luo country and western Kenya. Some foreigners deem movements like those just noted to be alien and exotic; but there are also in the area Christian churches, Catholic and Protestant, whose ways and creeds foreigners find more conventional or mainstream. As men and maybe women in Christian churches take upon themselves the privilege and duty of performing Eucharistic ritual on behalf of church adherents and the public, indigenous forms of sacrifice, including animal sacrifice, would seem threatened. Sacrificial prayer for Luo can become more individual or universalistic, no longer so much a matter of particular families, lineages, clans, or ogendni. No longer need elders remain the conduits of communication to spirits and divinity. But I doubt local sacrifice is about to die out completely, or that an individualist or universalist ethos will take over completely. Keeping multiple traditions going, even as these cross-fertilize, would seem to help keep faith in sacrifice alive by allowing each method to fail at times while another seems to succeed. Where this is so, believers need never question more basically the efficacy of sacrifice itself.

To assume, as many Luo seem to do, that lasting gain and satisfaction come only from loss and suffering, and that a painful death conversely can be a route to flourishing life, implies that martyrs will bring hope; and this is the sort of thinking that stories, historical reconstructions, like that of Alfayo Mango demonstrate. The deaths and expanded public legacies of countless other Nilotic prophets demonstrate much the same (Lienhardt 1961, Johnson 1994, Hutchinson 1996). But the principle applies not just to people who are, or become, ostensibly "religious" figures. It can also apply to martyred soldiers, politicians, and campaigners for causes of all sorts.

At the geographical and linguistic margins of the Kenyan nation, Luo have had their share of tragic and mysterious deaths of public figures (Schwartz 1995, Cohen and Atieno Odhiambo 2004). Most notably, Tom Mboya's death and Robert Ouko's, now internationally known, have been widely attributed to important persons who felt politically threatened by them or others acting on their behalf. But Luo at home and abroad can name many more than these. The great attention paid to victims of political violence in the press (when not under tight clampdown) and in colloquial discourse has been a tribute of a kind. The sometimes lurid details of, for instance, Ouko's half-burnt body (charred from the waist down, like Mango's) as popularly reported call to mind other kinds of burnings, animal burnings, more overtly sacrificial when they happen.

If it is true, as it seems, that animal sacrifice has become less frequent in recent decades in Luo country, the explanation may be not just that commercial butchers and marketers have taken the cattle or that churches have moved into the symbolic space. It is also that mass media have entered the communications, allowing everyone to participate after a fashion. The place to look for sacrifice is not just the cattle byre or sacred tree (*libaga*); now it is the church and the paper too. In time, assassinations and other murders become offerings—even self-offerings, since those who remember can call the person's decision to enter a dangerous line of work, or to adopt a risky position, a sacrifice of self to begin with.

If sacrificial idiom has a bearing on politics, it may bear too on economy, where the pain and gain of investment, and the bloodletting of corporate retrenchment, are common parlance, and where bankers lean so heavily on words like trust and fiduciary. To suppose that all this is mere simile, and that there is nothing culturally important behind this thinking to those concerned, would be too dismissive. For in some cultural contexts, it is precisely "the market" (if we are in it for our own financial or material gain) or "charity" (if we are in it for another's, or for our own political or spiritual gain) that *is* held sacred.

The perceived continuity between debts between humans, and debts between humans and divinity, appears in many of the sacred Judeo-Christian scriptures, but nowhere clearer than in these lines of the Lord's Prayer—to some Christians the holiest prayer, and one recited by Luo as Lamo Mar Ruoth:[22] "Thy kingdom come. Thy will be done in earth, as it is in heaven. / Give us this day our daily bread. And forgive us our debts, as we forgive our debtors" (Matthew 6:10–13).[23] And again, in the lines that follow the prayer: "For if ye forgive men their trespasses, your heavenly Father will also forgive you. But if ye forgive not men their trespasses, neither will your Father forgive you" (Matthew 6:14–15). These are Christian traditions that have become Luo and African people's traditions no less than anyone else's.

To make a sacrifice with a prayer—word and deed that give each other meaning—is to make an entrustment that reaches across the bounds between the known and unknown. Whether one considers these to be a form of bestowal, of communication, or just of self-gratification to reduce anxiety, prayer and sacrifice can be as solemn or sincere as any human act. They can be both instrumental and expressive. Sacrifice works to express what people cannot easily put into words, or to control that which cannot be controlled directly. Some feel it transports them into another state, bringing them nearer to divinity and each other. By communication through the death of the thing sacrificed, a sacrifice can represent an attempt to secure a piece of divine power, perfection, or favor, or to cleanse partakers of pollution: to procure the desired or get rid of the unwanted. Sacrifice can also represent time out of time: through its preparation, some feel, they enter a separate sphere, crossing a time horizon into a kind of black hole where the ordinary rules and sequences of living do not apply. Sacrifice, like other ritual, connects the present with the past and future, evoking what it does not directly denote or describe.

Luo and other Nilotes are far from unique in their conceptual webbing linking the entrustments of the living and the dead, the spiritual and terrestrial, and the sacred and profane. These are not just simple and direct cognitive equations but metaphors ritually expressed and symbolically charged or mediated. A creditor and a creator may not be deemed at all the same, yet in the minds of the indebted, their holds seem to share a certain likeness, one that also often evokes parenthood and patronage.

In conducting the sacrifices they do, Luo people partake of a broader tradition encompassing the Nilotic-speaking peoples whose homelands range from southern Sudan south into Tanzania, and including many of their Bantu-speaking neighbors as well. For all we know, aspects of this tradition may be millennia old. What changes are the contexts in which sacrifices are enacted and sacrificial ideas invoked. At present, in an era of HIV/AIDS epidemic, funerals have been so common in eastern equatorial Africa as to lay claim over many of the large and small animals that might otherwise have been available for other sacrificial occasions. And the threat of further impending illness and death has loomed so large as to preoccupy human minds, as if the struggle for health, however interpreted as a public or private concern, a medical or theological one, were sucking into itself all other purposes. The epidemic has not ended sacrifice, but it has sometimes given it new contextual framing.

The kinds of knowledge and belief discussed in this chapter might seem like pure faith to some, pure superstition to others. But it is their constant testing and their flexibility that keep them strong. Ancestors and other spirits, a supreme and all-encompassing divinity, and the un-personified order of the universe as manifest in kwer and chira all play different roles in Luo life; there is a rough division of labor between them. But when any one of them fails to provide answers in the human endeavor to explain, predict, and control, another seems to step in. In this way, if not in others, they help keep each other alive. The moral principles that Luo have tended to anchor in the ancestors, and that they see represented in divinity as revealed by ancestral visitations, contemporary prophets, the Judeo-Christian Bible, and daily events of many kinds, are principles of giving, sacrificing, and cooperating: principles with entrustment and obligation at their core.

CHAPTER 10

Conclusion

Entrustment and Obligation

Theoretically, payments of debt and executions of justice, when they cover the whole of social life, should tally. When they don't, there is something more to research and explain.
—MARY DOUGLAS

In these pages I have sought to describe a part of life that matters to people everywhere, and a part of Africa where it takes some distinctive forms. The basin of Africa's largest lake is not a place without access to credit and debt. Nor is it a land without deeper, broader experience of entrustment and obligation. The complex and subtle fiduciary culture already in operation, adapting in its own directions its rich mix of the old and new, and of the indigenous and exogenous—is a product of long experience and continuing ingenuity in Luo and East African culture. All of this is something that foreigners and officials may need to understand better before they land there—as they have continued to do for over half a century—with offers of new loans that they imagine will improve productivity or living conditions, educate the public in responsibility, or gain gratitude and political support. Learning about what occurs there already, and respecting the sophistication and dignity of the people concerned, is the way to start instead. Doing so is crucial if anything attempted there is to fit in, and it is what anyone concerned with Africa owes to African people. Taken seriously, the lessons of equatorial African experience—whether this be the cradle of humanity or not—might also benefit others elsewhere.

These pages have discussed tangible things, but ones with intangible dimensions deemed no less real for it. In suggesting what kinds of entrustment and obligation are found in an inland equatorial region of Africa still not well known beyond, I have sought to focus on items, thoughts, and actions regular enough to be important, maybe even defining, features of Luo and western Kenyan life. But many of these ways belong to no single ethnic group or tribe, let alone any "race," however defined. Customs and habits slide across linguistic boundaries. Languages and idioms traverse social ones. None of these need have much to do with race except in the ways people classify themselves and each other and set bounds for interacting.

Entrustment and obligation are important themes in many parts of the world, but it may also be that people in equatorial Africa have a special intuitive understanding about it. Some of this doubtless comes from the experience of first exploring the world as an object of entrustment oneself, riding around on a cousin's (or "sister's") back. Indeed, one of the first and biggest lessons about fiduciary life, and about human ability, is to be learned from children, especially girls. Watching them tending babies and toddlers, sometimes while farming family fields, teaches a remarkable lesson — if we did not happen to know it already from growing up in a large family stretched for supervisory care. It is that when given or allowed great responsibility over their juniors, many young people prove remarkably able and are sometimes honored and pleased to rise to it. A comparable lesson may be learned from watching boys entrusted with herds, sometimes on long loan to their elder kin from distant, unrelated adults. They learn from their peers as well as their elders. An equatorial African farming home is a school of responsibility.

Growing up in the lake basin provides other occasions for practicing and learning about entrustment throughout life: in animal lending and share contracting, in apartment sharing, and in marriages, funerals, and sacrifices. Being born with a set of debts and credits, and dying with another set to hand along, is a notion that might seem unfamiliar to one from outside East Africa who looks upon the start and finish of an individual life as the beginning and end of life itself. A life where final balance is never achieved, maybe not even desirable, can be an idea that may take some getting used to, but it is a realistic one.

Ever since Marcel Mauss's 1925 essay on the gift, anthropologists have been fascinated with that topic to the point where it has nearly eclipsed other forms of economic life, and of symbolic life with an economic dimension. "Gift economies" have been contrasted with "market economies" as though the pair defined the totality of economic life and the movement of life's broader flow. Portraits of tropical givers and sharers, like portraits of tropical traders and accumulators, have tended toward stereotypes that only too neatly fit the agendas of socialist and capitalist ideologues in capital cities and abroad, searching roots or outlets for their own preset agendas.

The Luo case in western Kenya suggests, however, that loans and entrustment, perhaps more than outright gifts or even sales, define the flow of transfers in an equatorial African setting. Perhaps they do so in many settings. Lending and entrusting are no less rich in symbolic meaning than is gift-giving, but often subtler. Gift-giving caught the attention of the anthropological world because it seemed at times to contradict the logic of the marketplace. Actors in the potlatch or kula converted wealth to power or influence, maintaining social structure and process, or expressing and reproducing valued norms, beliefs, and identities—rather than just piling it up. Borrowing and lending do all these no less; and it may be, as Mauss and his mentor Durkheim both suggested, that they are the process underlying gifting itself.

But whereas the gift-giving that Marcel Mauss studied (the potlatch, the kula, and so on) involved what he portrayed as obligations to give, receive, and reciprocate, this last aspect does not really capture the nature of fiduciary life, in East Africa or elsewhere. Just as important or more, Luo remind us, is the obligation to pass along: to continue a lineal or "vertical" flow. To Luo this can mean repaying the ancestors, satisfying their trust and perhaps the active interests of their spirits, by handing along their benefits to generations to come.

If entrustment, as enacted in East Africa, has a double nature—"lateral," or between the living, and "lineal," between the living and the nonliving (or the "living dead"), each of these can help sustain the other. In passing cows to a junior kinsman so he can marry, or money to a brother to pay his daughter's school fees, a senior kin satisfies not only those people and their living kin but also, by the same token, ancestors who are believed—nay, indeed known—to maintain an interest in lineage and kindred solidarity. One is also preempting the malicious wrath of a ghost of the young kin who might die prematurely—and take dreaded action not necessarily on that negligent person but perhaps instead on her own descendants. If the young man who marries, or the girl who gets to school, cannot or does not "repay" the favor, he or she may yet be expected, and feel obliged, to pass on a similar favor to a member of a generation descending.

Favors are to be repaid or passed along. So, think some, are disfavors. "Goods and services," the sunny side of economic and fiduciary life, should not blind us to "bads and disservices." As Luo speak of them, homicides, injuries, and neglect are all repayable, or alternately, they may all be passed along, rather like toxic waste. Vengeance and aggression seek their own outlets, not necessarily in the direction from which harm has come. None of this means that the plusses and minuses in life, however defined, must cancel each other out.

In transfers between the living, just as between the dead, the living, and the unborn, the possibility of lasting asymmetry always remains a real one. Even if borrower and lender both attempt to square their "accounts," the changes in relative values affecting what is borrowed and lent, or entrusted, over the interval makes an even reckoning both difficult and unlikely. Moreover, lenders often have good reasons not to have their loans repaid. Outstanding debts are not just a kind of wealth but also a kind of power. More basically, a loan or entrustment (of a cow or goat, for instance) can express trust, constituting a kind of social circuitry as kinetic as electricity. In Africa or elsewhere, a life in which all debts were settled would be a frozen life of atomized individuals—no life at all.

Entrustment is a big topic. It is complex, morally weighty, and hard to define. And yet there are ways to simplify it (something true of any complex topic) and to explore it directly, with tangible evidence. Credit and debt, like the xylem and phloem of a living tree, take more forms and follow more roots and branches, more capillaries of cause and effect, than anyone has traced. In this volume I hope to have sketched a cultural cross-section of that plant, shown how it grows in a particular African setting, and described some of its subjective dimensions. I suggest it to be more deeply rooted and socially entangled than many financiers suspect when introducing loan programs from outside, and to involve longer lasting imbalances and asymmetries than many ever dream. Credit and debt are not new ideas in Luoland, despite the near absence of specialized moneylenders in the rural areas. Luo borrow and lend among themselves for production and consumption, in cash and kind, over the short term and the long. Loans and entrustments involve not just middle-aged men but women too; and the young and old; and the dead and unborn. They appear as part of "multiplex" social bonds—they accompany kinship, friendship, church membership, commercial custom, and so on, which may coincide—and some local lenders depend precisely on these overlapping ties for their repayments. Loans are the elastic in economic life, stretching labor and capital over land, stretching food over the season, stretching income over a lifetime. But entrustments—loans with feelings behind them, and ones not necessarily repayable directly—are not just about wealth, or about giving and getting. They are also about defining who we are, and about connecting to something bigger than ourselves.

Entrustment and Obligation

Economic life is cultural too. This exploration of borrowing and lending in western Kenya has shown that loans and repayments are shaped as much by norms, beliefs, and social organization as by supplies, demands, and prices. It has suggested that insiders' views can differ sharply from outsiders' and proposed that qualitative and relativistic interpretations balance quantitative and absolutist ones.

Contrary to textbook wisdom about economic evolution from barter to cash to credit—a schema that always seems to place Europe or North America at the top—Luo live in a credit economy already. It seems they always have. But the great diversity of loans and debts, and their often profoundly embedded cultural meanings, are hardly captured by an economistic term like "credit," connoting currency, written agreements, and durations limited to the human life span. It is often fitting to use more inclusive terms like *entrustment* and *obligation*. Luo economy has long been webbed with credits and debts, entrustments and obligations, before and beyond those brought in from outside their country.

Just about anything, in western Kenya, is lent or entrusted: land, money, seeds, radios, livestock, labor, and even humans themselves. It is not just banks, cooperatives, and merchants who lend: it is also relatives, neighbors, friends, and church mates. Most loans and entrustments are not monetary. Twentieth-century currencies have interwoven with other means of exchange, in loans as in immediate sales, but they are not expunging them. Financial entrustment in rural East Africa is seldom neatly separable from other economic entrustment. Luo borrow and lend in many ways beyond those normally classed as "credit": for instance, in rotating farm labor, in delayed marriage payments, in contributions to labor migrants in the expectation of later remittances, and in the fostering and pledging of children. Like other East Africans, Luo do more of their borrowing and lending unofficially than officially.

Who is entrusted is not always obvious. Even in the simplest case, a loan between two individuals, a lender places his or her trust not just in the borrower but also in that person's upbringers and socializers, in mutual acquaintances, in other prospective borrowers, or in society at large to ensure the return. Borrowers, for their part, make their decisions to borrow and repay partly on the basis of who else is able to make claims on resources once borrowed, and who else around will take note if they default. If only subtly, loans always involve third parties.

Loans and entrustments tie tightly into other kinds of exchanges. Lending can be a way of saving. It can help lock one's assets into a form that will not be taxed, devaluated, begged away, or devoured by weevils. In some ways distinctions between the two terms are artificial, as when participants in rotating credit associations both lend and save in the same transactions. In rural Africa neither credit nor savings can be well understood without understanding the other.

Contrary to some economists' assumption of a "liquidity preference," farming people try to keep a good part of their property illiquid, removed from their own temptation. Insofar as their complex feelings can be reduced to "preferences," theirs is just as much an "illiquidity preference." They want to have a good part of their wealth accessible in a crisis but not just freely available all the time to dip into or be dipped into by others. Lending is one way. Keeping the wealth tied up in indivisible forms is another. Animals, trees, farm tools, iron roof sheeting, music machines . . . all are savings, among other things; and all are potentially saleable in a crisis. And as every Luo knows, none of these burns a hole in the pocket in the way money does. Most saving, like most lending, is nonmonetary, for good reasons — including money's divisibility, its contestability, its inflation, and its manipulability by government.

Lending can also be investing, as when a livestock borrower offers better grazing land than the lender has available. Lending can serve as a kind of insurance, as when the livestock lender spreads his herd out beyond the reach of any single stock thief or any likely local epizootic. It can be a kind of social security, as when school fee payments to a promising youth help guarantee the benefactor a more comfortable old age in return. Lending can shelter one's wealth from taxes or chiefly demands for bribes or tributes, since it helps keep one's wealth inconspicuous. Through their loans Luo, like other East African people, seem to try to keep their wealth both in diverse forms and in diverse hands.

Credits and debts also tie closely into many kinds of interfamilial transfers, share contracting agreements, and remittances between city and countryside. Loans in rural East Africa do not divide neatly into sorts for consumption, production, or investment, as in conventional economic parlance. (How, for instance, would one class the loan of an ox used for plowing, slaughtered and eaten in a funeral, and repaid with another animal?)

That terms like *credit, saving, investment,* and *income* often cannot translate directly into African languages suggests they are often perceived as interconnected—or to put it differently, not cognitively separated in those ways to begin with. Loans connect serially with gifts, sales, share contracts, and thefts. They may be indistinguishable from these or be reinterpreted over time from one category to another, as when livestock loans to agnatic kinsmen or land loans to jodak eventually become permanent transfers. Loans between unrelated friends can turn into devolution among kin as kinship and genealogies themselves get dismantled and re-welded over time. All these categories reconfigure as one moves from one culture to another.

Luo incur debts and obligations not just actively, but also passively. They do so not just by individual contract but also by ascribed membership in groups, networks, and categories. Luo jodak are born with obligations to the weg lowo who have invited their forebears onto the land. Luo men grow up to inherit claims or liabilities for marriage payments, or they are persuaded to share in these for fear of looking antisocial. By legal means or "supernatural" sanctions like ghostly vengeance, they are held responsible for damages to people and property incurred by their kin. What they owe or command depends as much on who they are—and thus on whom they are related to—as on what they have done. Debts actively acquired mingle with others passively acquired, as in the case of a jadak who compounded his kin group's indebtedness by murdering a wuon lowo.

Accordingly, small-scale farmers in Luoland incur entrustments and obligations over very short and very long periods, up to generations. Loan terms depend on what is lent. While no one borrows a plow for a decade or generation, a land or livestock loan can last two or three generations, partly because these belong to groups, networks, and categories as well as to individuals. No borrower is a tabula rasa. Everyone always has debts or obligations to pay and to collect, and resources from one part of this credit economy can easily get shifted right into another. While borrowers sometimes compete for loans, creditors must compete for repayments too.

Credit, Debt, and "Spheres of Exchange"

Three of the questions underlying this study have been: What is exchangeable for what, why, and does the passage of time make any difference? Do Luo perceive in their economic lives anything like the anthropologists' "spheres of exchange"? Or are all resources, as in some economists' and anthropologists' assumption, ultimately interchangeable? Care has been taken to distinguish between a people's ideals (the "ought") and their real behavior (the "is"), since these often differ.

When interviewed in depth, rural-dwelling Luo people do *not* speak or behave as though everything ought properly to be exchangeable for everything else. Wealth to them instead seems something like a maze with many walls and one-way turnstiles, or to use a different simile, a circuit with complex pathways of switching. Converting grain to animals is deemed better than converting animals to grain, except in desperation. The loan of a male goat is expected not to be repaid with a female, even though their cash values may be the same. Humans are not to be bought and sold in the marketplace. To sell patrimonial land is to some an even greater infraction of custom still. Wealth acquired by means too easy or antisocial, including *pesa makech* or "bitter money," is expected to be kept separate from other wealth. To breach these categories calls for expensive and possibly dangerous ritual purification. Yet in reality many Luo do in fact breach these categories, directly or indirectly through "laundered" transactions.

How the gates or switches are set depends on who one is. Elder men and women seem to care more about keeping "bitter" money discrete than juniors do. Elder men like to see wealth converted from cash, grain, sheep, or goats into cattle, but women and juniors are not so clear about it and may even be opposed to such conversion. Foreign bankers in the Luo country like to see their debtors convert their herds into liquid cash—and thus tend to disagree with lineage elders. Animate wealth may not be commensurable with inanimate; prestige property with subsistence property, ephemeral with permanent; and such distinctions all help determine what is properly exchangeable for what. Some Luo people clearly expect borrowed wealth to be used differently from earned wealth. Some have insisted to me that borrowed money is not makech, since the repayment will need to be earned; but others disagree, saying it is like a windfall.

Lending terms do not necessarily ride on the cash values of things lent. Neighboring farmers devising share contracts with plows, oxen, and land sometimes choose to overlook the unequal values of these in a fiction of equality that creates social symmetry between co-lending partners. In other instances, fictions of equality conceal uneven exchanges, as when a plow-team owner arranges four days of plowing labor by a neighbor at the peak plowing season in exchange for four days' use of the plow team when the optimum plowing time has passed.

The time lag involved in a loan can affect what is exchangeable for what. The buffer of time makes a locally unacceptable transfer, such as the exchange of rights in a daughter against grain during a famine, more acceptable to Luo in the form of a pledge, with a contingency condition and a time lag built in. Things symbolically incomparable, like money and cattle in Luoland, can still be economically interchangeable or substitutable (to economists, "fungible") in delayed reciprocity or in exchanges brokered through third parties or mediated through other goods and services.

Yet at another level, these diverse loans are economically comparable, and competitive, both at borrowing time and at repayment time. Human labor is exchangeable for land, livestock, fertilizer, or radios; and all are exchangeable in one way or another for money. Spheres of exchange, even if conceptually distinct, link up by indirect or "laundered" exchanges. A loan of fertilizer and pesticides is convertible (and indeed, commonly converted) to a heifer for bridewealth: a short-term exchange cycle feeds into a long-term one. This is usually a one-way conversion. Bankers and cooperatives officers who expect their chemical loans to be repaid from farmers' livestock herds are naïve in a cultural sense, while perhaps quite reasonable in some narrow economic sense.

In sum, the cognitive and behavioral barriers that at first blush suggest "spheres of exchange" are hardly simple or watertight, in local theory or in local practice, but they do stand as strong predilections, and as strong warnings to Luo and to those who borrow and lend among them. The symbolic and economic realms cannot be captured by separate disciplines. Whether exchanges between the spheres—between the local and exogenous, or between the sacred and profane—involve moral transgression is hotly debatable in cultural terms. And whether they should somehow equilibrate by supply and demand, for efficiency, is debatable in economic terms. Luo, like other people, can think more than one way about these things. To understand the cross-currents means comparing insiders' and outsiders' points of view, without stereotyping either. In the study of economic culture much comparison remains to be done.

Intimacy and Fiduciary Life

Lending and borrowing, and the entrustments and obligations underlying them, hinge on identity and intimacy. What you may borrow, how long you may keep it, and whether you must put down collateral or pay interest usually depend on who you are and who the lender is. They depend, to begin with, on your ethnicity, age, gender, and class—things understood not just in absolute terms but also relative ones. They depend on how closely you and the lender deem yourselves to be related, and how: by kinship, friendship, or neighborhood, by previous trade dealings, by a school tie, or perhaps by co-worship or an introduction by a mutual acquaintance. Groups, networks, and categories all come into play in determining the nature of social distance. The tie helps determine the nature of the loan, and the loan in turn helps define the tie.

Luo people normally try to give their closest contacts (as they define them) the easiest loan terms, and strangers the worst. Pressures to do so can, however, discourage them from lending in the first place. Many hesitate to lend to kin particularly where money is concerned because they fear having to put the screws on a relative for the return and being thought antisocial.

Luo are no strangers to profitable lending, but they do not take "interest," or loan profit, for granted as a natural growth of capital. Interest is uncommon in familial or friendly loans, and it is far from universal in loans between strangers—for instance, between Luo and Arab or Indian minority traders living among them. Moreover, even where "interest" is agreed upon, it does not keep accruing indefinitely if a debt goes unrepaid: after a time an unrepaid loan is normally left as dead, and future credit just denied. A cash loan repaid at par, in an inflationary era, conceals a hidden transfer to the borrower, since what is repaid is worth less than what was lent. Seasonal fluctuations in values can be more important still: who benefits, and how, always depends on when in the yearly round the loan is made and collected.

Loans and debts, as already noted, concern not only borrowers and lenders but also others with interests or claims on their resources. In communities with much face-to-face interaction between people with multi-stranded social ties, and little real secrecy about wealth, these social considerations become supremely important. Poverty seems to sharpen them. An individual's decision to save, borrow, or lend takes into account not only his or her own needs and capabilities but also the pressing demands of kin, neighbors, and others using the same resources. It thus must take account of general seasonal supplies and demands. Selling grain cheaply after harvest and buying it back expensively in the lean season—the subtle "pawning" practice that is a recurring pattern of Luo marketing—may look shortsighted economically if one considers only the prices concerned. But it can unload one's storage risks onto the shoulders of the buyer. And from a social perspective, too, it can be the prudent thing for one to do, since, as we have seen, selling liquid wealth to buy illiquid wealth can help defend it from the constant claims of kin and neighbors. Similarly, contributing to a rotating credit and saving association as a member who will not collect until late in the group's cycle means receiving negative real interest, once inflation is taken into account. But here again, the practice affords the member an honorable, sociable-seeming way to say no to the daily demands of friends and relatives on his or her cash. Delayed exchanges that seem irrational by market standards sometimes conceal hidden reason.

Customs and beliefs that look egalitarian in one way are often inegalitarian in another. Luo people often call themselves egalitarian, and are called so by others. Their land loans help even out imbalanced personal or familial holdings, allowing land to the needy and most capable and helping ensure that most anyone who wants land to farm can get some. But at the same time these loans engender important status distinctions that can last longer than a lifetime.

Conversely, hierarchy can imply its own subtle equalities. A segmentary lineage ideology and gerontocracy impose heavy social duties on young men, but they also guarantee each man a chance to become a patriarch and, after death, a lineage and clan head. Young men who must hand over a large part of their cash earnings to their fathers with no questions asked, or who are demanded to supply easy grain loans to their poorer uncles, can expect to enjoy their own turn in time. Junior wives who seem much put upon by senior wives can still hope for yet more junior wives or daughters-in-law to answer to their own orders. Women who seem perpetually edged out of wealth by virtue of their gender or poor marriages can nonetheless command respect as sages, diviners, healers, possible witches, or advisers to politicians, in their old age. Junior kin who neglect their needs, refusing them sharing or charity, are taught to fear ghostly vengeance after those elders' death.

Symbol, Sacrifice, and the Flow of Trust

Economic entrustment is not neatly distinguishable from ritual, symbolic, or spiritual entrustment. As the multisided symbolism of the roof spire stick (osuri), sometimes forbidden for jodak, suggests, borrowing and lending can place people on an unequal footing in ways not just economic, involving innuendos of sexuality and seniority.[1] Many of the biggest economic transactions many rural Africans may make (like people in many other parts of the world) are for life-cycle crises: births; circumcisions or initiations, where practiced; and marriages. Entrustment and obligation involve the dead as well as the living. Marriage transfers of cattle and other things, often protracted over decades and even over generations, are part of the most basic stitching holding society together, among Luo as among their Nilotic and Bantu-speaking neighbors. These involve not just bride and groom, as seen, but numerous kin expected to contribute and to receive in chainlike webs on both sides of a marriage. While the bulk of marriage payments goes from groom's side to bride's, not all the transfers flow in this direction; a sequenced back-and-forth is part of expected procedure. That marriage transfers are returnable in the event of divorce or a married woman's untimely death underlines the conditional and provisional quality of marriage; and the protraction of payments, and deductions made from returns, show that claims in individual "property" are seldom clear cut even though herds are controlled by individuals. Unmarried sex has its own fiduciary side.

Entrustment plays a vital part in funerals and burials too. Contributions to burial societies can use large parts of personal money income. Funerals are among the largest and most important gatherings in Luo life. In an East African way of thinking one may understand funerals in part as satisfying the dead by assembling the living and keeping their exchanges active, and some kinds of inheritance as repaying debts to the dead by passing on property to the unborn, or by seeing to its fair distribution among kin and heirs: a very difficult process fraught with possibilities of disharmony and schism. Social relations—kinship and ethnicity, gender and age, class and religion— always influence real resource allocation. And as East African people understand things, often the social relations include persons not present or powers unseen.

Entrustment is a main rationale of sacrifice to ancestral and other spirits, and to divinity. Sacrifice is a complex topic. It is hard to find any single common element in all things called *liswa, misango, dolo,* or *msiro* in Luo, or "sacrifice" in English. They take many forms, serve many ostensible purposes, address themselves to different spiritual agencies and impersonal powers. There are, however, a few familiar, ever-recurring motifs in Luo and East African ritual repertoire, and discernible patterns in sacrificial purposes and procedures. Indigenous understandings of sacrifice have blended together with understandings from Judeo-Christian tradition, particularly where these concern suffering and salvation. People in western Kenya and around East Africa speak of sacrifice—that which they conduct themselves or that conducted on their behalf—in an economic idiom that reflects their own worldly dealings among themselves, but in doing so they stretch that idiom. Bargains, contracts, and covenants figure importantly, but neither a legal nor an economic idiom captures sacrifice entirely or roundly reflects the symbolic and emotional content of the entrustments and obligations involved, or the involvement of third parties seen or unseen.

If sacrifice involves a visible flow of blood, smoke, or wine, it also implies a flow of trust. Ideas about suffering and substitution, and the ability of one being to stand in for countless others, all play a part in these understandings. Metaphor blurs the boundaries between being and resembling, and metonym those between the part and the whole, as ordinary understandings are suspended during ritual enactments. In common parlance or in concerted action of protest movements and churches, the symbolism of sacrifice imbues not just religious but also political and economic realms, and where these things combine, emotions can run to extremes. The periodic resurgence of protest movements in East Africa shows how little understood and unpredictable these things can be.

Now some further mysteries and unknowns. While entrustment and obligation lead across the bounds of theology and politics, they also lead into questions of prehistory, psychology, and ethology that might seem more scientific in a way but are still shrouded in mystery. Whereas some classic evolutionary theories have presumed a general human movement from barter to cash to credit economies, as seen, evidence from East Africa and elsewhere suggests these have been too presumptuous. There is no sign that anyone there ever relied less on credit, broadly defined as borrowing and lending, than in our times. This does not mean that all social evolutionary theory is irrelevant to Luo, western Kenya, or East Africa. But it does mean that the questions must be recast in less-loaded, ethnocentric terms, ones less conducive to painting human deficiencies in tropical places.

A flatter, fairer line of inquiry might be the following. Humans borrow and lend, and they make entrustments in serial forms like inheritance too. Kinship and familiarity can breed distrust, but they tilt the pitch toward trust instead. What evidence we have suggests that Luo and other East African people invest more trust in their kith and kin than in strangers, and honor such trust more faithfully themselves. If this is so, then might such a pattern be construed as one instance of a general principle of biological or social adaptation? Do credit and entrustment make us more fit to survive?

The question is hard to answer, but no less important for it. Human and other animal evolution produces propensities not just to compete but also to cooperate—within species as well as between—and cooperation is about the best evidence of anything resembling trust. The ability to make decisions about whether to cooperate or not, and to decide according to shifting conditions, is an adaptive quality of humans and at least some other animals.

To put these things within an evolutionary framework (for instance, a Darwinian one) need not mean interpreting all thought and action as genetically programmed in a rigid way, or as pertaining only to individual organisms and their own offspring. Rather, the ability to change and adapt thought and action may derive from co-adaptation of different genes, and from a broader co-adaptation between genetic endowment, learned experience, and intuitive or other thought, whether this thought be traced to individual minds or shared culture (Bateson 1988, Deacon 1997). If humans have instincts, these need not be just "hard wired" or just acquired through communication and experience, but they can be something of both. If genetic endowments manifest themselves in human thoughts and actions, it seems, they do so not as simple causes but also get reset like dials or switches being reset by environmental and social influences—in short, by lived experience. Any such process is probably an iterative, that is a recursive one.

Cooperation may result from evolutionary process in any of several ways, some still hypothetical. First, aid to an individual organism's relatives, if it increases their chances of survival, may make the very propensity to aid one's relatives (or those one perceives as one's relatives) more likely to recur in subsequent generations. Second, groups or networks of individuals may, by cooperating, gain an edge of some kind over other groups. "Natural selection" can apply to groups as well as to individuals; and ironically, by cooperating internally, some groups can out-compete other groups. Third, even if two individual organisms (say, within a species) are not related, they may both, by cooperating, increase their chances of surviving and reproducing.[2] In humans and other animals, a growing familiarity between members of a group is one of the conditions that can induce each to trust other members, even though I would add that it might lead to distrust, in addition or instead. Self-awareness, including the ability to predict how another will respond to one's own actions, may be essential to the trust on which cooperation is based; and an ethologist might construe self-awareness itself to be a product of evolution.

Evolutionary theory about cooperation and trust has not yet been well connected to empirical findings about economic entrustment and obligation, between humans themselves or between humans and the spiritual beings of whom or of which they claim to know. A few caveats are in order. Further enquiries along these lines should accept, for a start, that different kinds of human groups both cooperate and compete at once. Ethnic groups, kin groups, social classes, political regimes, and so on may all be analyzed this way. But one must not confuse individual with collective purpose. Nor must one confuse motives with results. The motives by which a group's members operate need not be the same as the outcome of their collective thoughts and actions. Then there are problems merely in gathering data. A hard part of studying social evolution is proving that particular kinds of cooperators (or competitors) outlive *or* out-reproduce other individuals or groups. Just gathering firsthand evidence might take many lifetimes. There is also political hazard to take into account. Studies that consider humans and animals together in an evolutionary framework will be hard enough to keep free of unfair value judgments or the appearance of them. Racism and ethnocentrism have a way of creeping in through the windows and becoming an issue even when locked out.

Intimates and Strangers

Ethnography need not contain a moral like a story, but one who wishes can certainly find basic moral lessons to be drawn from Luo and other East African experience as I have endeavored to represent it. Here are just three of them. Lessons like these can be derived by interpreting a story like the one a Luo *pim*, the grandmotherly nursery instructor, uses to trace these people's origins, the story of the lost spear and the lost bead, briefly recounted in Chapter 3. I hope to have shown that they are discernible behind a variety of contemporary practices too.

The first concerns cooperation and reciprocity. Humans must be willing to give and take, and that includes being taken from when the occasion justifies. They cannot get by without each other's help and forgiveness. Reciprocity is crucial, *but* possessiveness with reciprocity too strictly enforced—as with the spear and bead—can lead to ruin. Loan can lose itself and friend, but even among friends and relatives in small circles we humans will continue, as we always have, to become creditors and debtors before we have had time to look around. People will forever be in each other's debt, and it is best to accept it.

A second lesson is that intimacy means potential rivalry too, and that special care must be taken to make sure that rivalry is contained. Brotherly love risks turning as bitter as any, as bitter as any of the worst between co-wives where found. Losing a possession, or an accustomed claim of access, can sting more than acquiring one can please. Betrayal of trust, too, can spoil a relationship more quickly than faithful reciprocity can build it. In ways like these, human emotions are often lopsided. But life and solidarity are more important than any piece of property. Matters of intergenerational conveyance and inheritance, where the belongings gain extra meaning by symbolizing the elderly or dead themselves, can be some of the most trying. Here we may need to rise to our most forgiving.

Third, our contemporaries are not the only ones who matter. Even if one's only concern in the end is the thoughts and feelings of those currently living and breathing, reference to someone or something else can be of value. Human decency and fairness are hard enough to maintain anywhere, let alone where spaces are confined, material and economic resources scarce, population dense, and competition therefore likely. One way of keeping peace is to remind each other of some watchful eye connected with the past or future, or in a timeless condition. Some turn to ancestors, some to the future ghost of a mistreated kin, some to a supreme creator left to judge the world. Luo and Kenyan people variously turn to them *all*, and to an impersonal force of nature besides. The recurrence of ideas like these across the lake basin (and across so many cultures around the tropics and the world) suggests that the mere believing or perceiving, and maybe the discussing and agreeing, may have some adaptive value in itself — *whether or not* the thing referred to is materially real or true.

The primacy of reproduction and continuity as a human concern comes out nowhere more clearly than in eastern equatorial Africa, where the Luo word for doorway also means descent group and one can hardly pass through one without thinking the other. Here everyone has a duty, at least to the near and dear, to see that they too be able to mature as moral people, to marry, and to procreate — to carry on a flow. What we receive from those who have gone before we cannot always give them back, but must sometimes just pass along. Attention to the before and after affords a way of belonging to something lasting, and of transcending whatever troubles of the time one cannot solve.[3]

From time to time we have seen evidence that entrustment and obligation exercise power and shift it. Entrusting children and livestock cements alliances not just between individuals but also between kin groups or communities. Credit, like free charity, can mean patronage, an asymmetrical tie of power or influence. Lending displays strength; borrowing, weakness. Long-term entrustments between relatives or friends, as in school fee contributions or bridewealth contributions, give the provider a certain moral ascendancy over the recipient.

Some of what holds in small circles also holds in large. Among not just intimates but strangers too, as the companion studies will show, attempts to control wealth can mean attempts to control people. Institutional lenders can lend to extend patronage, incurring political obligations, and they sometimes get their arms twisted to lend by the more powerful still. Loans of cash, from cooperative unions, from the capital city, or from the World Bank, are the stuff of patronage and clientage as well as of potential aid. At all levels of scale, they imply unequal status and asymmetric influence. They can be ways of spreading influence over people at the edge of markets, or of spreading favors and wooing voters and supporters. But they can also damage dignity and incur resentment that verges easily into anger.

If local fiduciary affairs involve notions of morality and ethics — bringing both the temporal and spiritual to bear and blurring the difference — so too can long-distance ones. When official lenders speak of an agricultural "sector" and of intersectoral "leakages" — for instance, when their borrowers use "farm loans" for school fees, medicines, or nights on the town — the lenders invoke their own ideas about what anthropologists have called "spheres of exchange." When aid agency loans get turned over to animals for bridewealth or sacrifice, when mortgage lenders and "repossession" agents threaten ancestral grave sites, or when people turn to divinity for protection against these people called "raiders" — just like cattle raiders from across a border — communication channels are challenged. People experienced and maybe self-educated as farmers, herders, fishers, and migrant workers, and others trained as bankers and lawyers, then search for a common vocabulary. Finding one might bring them together in a way but also threaten to force them into uncomfortable encounters with each other's philosophical and spiritual lives. Insider and outsider find a challenge of epistemological interpretation, let alone of brokering cultures.

Luo and Kenyan people understand borrowing and lending; no one does better. But in the past they did theirs mostly locally, or among familiars. National and international credit programs, explored in the companion studies forthcoming, complicate a picture already complex. They bring not just new seeds but also new language into the picture. When loans and debts bridge cultures, and complete strangers find their fortunes entangled, the meanings of entrustment and obligation spring into question. Beliefs on all sides come open to challenge. Norms, values, and definitions become even easier to manipulate, and social distance and responsibility more debatable. Richer and richer meet poorer and poorer, the juxtapositions becoming so extreme they challenge the very definitions of wealth and poverty. Borrowers, lenders, and brokers between them find themselves involved in explosive issues of cultural integrity, nationhood, and sovereignty. The door for misunderstandings swings wide open.

Notes

Chapter 1: Introduction

Epigraph. Wittgenstein (1984, 74). He was referring here to conversation, and elsewhere by the same metaphor, to senses of humor, and fashion, shared or not shared.

1. Prefixes of ethnic group names and languages are omitted in this study except where necessary for clarity. In the Luo tongue, properly Dholuo, a single Luo is a Jaluo, plural Jo-luo. Sometimes the root is capitalized, as in DhoLuo, to distinguish prefix from root.

2. In this work names of neighborhoods and individual interlocutors have been changed (except in the acknowledgments or where individuals asked to be named) to safeguard confidentiality. The names of larger administrative locations and the ones on which these were based (*ogendni*, discussed later) before some were subdivided have been left as generally known.

3. G. A. S. Northcote, District Commissioner, South Kavirondo, Annual Report. His heading 349/314/18 (page unnumbered). Kenya National Archives, Syracuse Microfilms, Film 2801, Reel 37. Hereinafter, the repository and collection are abbreviated "KNA" and "Syr." Northcote, an Oxford graduate who had learned to speak Luo in Karungu from 1904 and then moved to Kisii in 1907 to set up an administrative station for South Kavirondo (later South Nyanza), was known for his moderate, conciliatory approach through a tense time of popular Gusii rebellion and harsh British military reprisals. Ironically, eleven months before submitting this report, he had taken a spear in the back, as a conspicuous symbol of the new colonial presence, while British hard-line military commanders who ordered a machine-gun slaughter of spear-armed Gusii went unharmed (see Wipper 1977, ch. 2, Maxon 1989, ch. 2). Northcote later become governor of Hong Kong.

4. For the present, here are a few basic definitions for terms with economic usages; some of these need additional definitions or multiple translations later. A *transfer* refers to a shift or conveyance of ownership, possession, control, or other rights or duties, between persons

or at least *from* or *to* one or more persons—with respect to an animate or inanimate *good or service*. (It may also refer to a shift in a *"bad"* or *disservice*—the more neglected but equally important side of economics.) *Exchange*, a term that often carries more general or abstract meanings, usually implies a reciprocation of some kind, direct or indirect. A *loan* or *credit* is an exchange expected to be reciprocated after a lapse of time. It implies a *debt*. A loan may occur between one person or group and another, or between a group and any of its members. It may be returnable in like form or otherwise. It may or may not involve a nominal or real gain or loss (that is, positive or negative "interest"). A transfer or exchange involving this or other faith, trust, or confidence is *fiduciary*, a term often applied to money or money-like devices but one with much broader applicability.

5. In this sense entrustment comes close to the original Latinate meaning of credit, from *credere*, to believe or entrust.

6. The phrase "symbolic capital" comes from Pierre Bourdieu. He writes, "Symbolic capital, a transformed and thereby *disguised* form of physical 'economic capital' produces its proper effect only inasmuch, as it conceals the fact that it originates in 'material' forms of capital which are also, in the last analysis, the source of its effects" (Bourdieu 1977, 183).

Chapter 2: Fiduciary Culture

Epigraph. Mauss (1967, 1.1).

1. Boas wrote in his *12th Report* of 1898, "The economic system of the Indians of British Columbia is largely based on credit, just as much as that of civilized communities. . . . The contracting of debts, on the one hand, and the paying of debts, on the other, is the potlatch." Quoted in Mauss (1967, 100n).

2. DuBois on Alor recorded elaborate verses about finance—dramatizing, with all-night dances and gong-playing, the roles, yearnings, and fantasies of what she called creditors and debtors in an involved system of ceremonial and other prestations. She called this versification "practically the only attempt at literary expression of any sort" among Alorese (1969, 136). Ancestral spirits were recruited to help in dunning debtors (pp. 137–43).

3. For this Mauss has been criticized by James Carrier and others. Eighteenth-century social contract theory played a part in Mauss's thinking (see Sahlins 1974). But whereas Rousseau (like Hobbes and Locke before him) had conceived of the state of nature, and the natural man, as having no social contract, Mauss looked nostalgically to the primitive—or the nearest living approximations—to find a happy state of solidarity.

4. American anthropologist Annette Weiner, pulling apart Malinowski's and Mauss's male-centered, reciprocity-focused treatment of Trobriand and Melanesian exchange, emphasizes what she calls keeping-while-giving (in cloth wealth, for instance, or in heritable intangibles), as a way of understanding female power, ranking, and cultural reproduction in the Pacific. Others who have dissected and reworked Mauss's writings on Polynesian and Malanesian Pacific island exchange networks include Claude Lévi-Strauss, Maurice Godelier, Marilyn Strathern, and Nicholas Thomas, to name only a few.

5. Testart (1998, esp. 103–4). See also Falola and Lovejoy (1994), and Lovejoy and Falola (2003), on precolonial, mostly western, African cases where people pledged themselves or kin as collateral against loans, with sanction of slavery for default. In western Kenya pledges of

humans, for example, of a daughter in a famine for marriage, need not imply slavery. Elsewhere, in the case of charitable donations anywhere they occur, donors may of course expect or require some sort of "return" like thanking, or naming a building, in addition to their own satisfaction in giving.

6. This is not to deny that Japan has its distinct ethnic minorities—for instance, Ainu—and underclasses such as the Burakumin.

7. Postwar Japanese corporations capitalized on fictive kinship, instilling in their younger workers a feeling of obligation and loyalty explicitly modeled after Japanese-style family units (*ie* and *dozoku*).

8. LeClair and Schneider's edited collection on economic anthropology (1968) offers articles clearly representing both formalist and substantivist sides of the debate.

9. Trobriand kula and northwest coast potlatching may, however, involve magic to soften up one's giver. And the same people who participate in these ceremonies with all their gifting can also participate separately in market exchanges with haggling and other forms of what Sahlins calls "negative reciprocity."

10. Arjun Appadurai and contributors to his volume (1986) have noted the same in other settings.

11. Note the parallel in the exchange of meat from large animal kills among hunting people. Typically, the kin or co-residents of successful hunters give away most of the meat from a large kill, in at least tacit expectation of receiving some back when their recipients make one. This works well, and may be quite necessary, among mobile people for whom food storage by salting, smoking, or refrigeration are impractical or impossible.

12. The "méconnaissance," or misrecognition, Bourdieu described recalls Marx's idea of "false consciousness," but he offered no explicitly revolutionary program like Marx's. See also Parkin (1972) on coastal Kenya, and Glazier 1985 (on Embu in central Kenya), on the cultural masking of self-interest in African land dealings.

13. More is said about all these issues of "development" finance, bureaucracy, and statehood in the volume to come on credit between cultures.

14. In this discussion it helps to remember that what *is* trust and what is *called* trust are not necessarily the same question: as Ludwig Wittgenstein never tired of insisting, words have the meanings we give them.

15. Hart's view of trust here brings to mind Clifford Geertz's in his study of the rotating saving and credit association: as a "middle rung" in development (1962). Along other lines, compare Hart's understanding of urban trust with Abner Cohen's on the ethnic-specific credit ties of Hausa cattle and kola nut traders in Yoruba towns in Nigeria (1969).

16. Ethnography about economy in societies with collapsed or severely dysfunctional nation states is still in its youth, but progressing quickly. More economic life can be found in some of these settings than most outsiders ever imagine: for instance, in Somalia's cross-border cattle trade and "informal" currency markets (Little 2003). The role of trust and "social capital" in these settings is a rich field for further social and cultural research.

17. Elsewhere (Shipton 1994) I discuss the numerous strategies of rural Gambians for borrowing and repaying different quantities despite the Islamic prohibition on "interest." Lenders commonly insist on switching the good or service between loan and repayment, partly to conceal an increase in value demanded or make it hard to measure. See also Stiansen

and Guyer (1999a and b) on Islam and credit elsewhere in Africa. Much further research remains to be done on the topic.

18. On the manipulation of economic "custom" for interested purposes, in land acquisition, for instance, see Parkin (1972) on Giriama, Glazier (1985) on Embu, and Moore (1986) on Chagga. Ostrom and Walker (2003) round up experiments with trust games.

19. See Guyer (1981), Netting et al. (1984), Moock (1986), and Guyer and Peters (1987) on uses and misuses of "household," and Guyer (1981), Berry (1984) and Robertson (1984) on "community."

Chapter 3: Luo and Their Livelihood

Epigraph. Cohen and Atieno Odhiambo (1989, 27), on Luo in Siaya District, an area then larger than now. Obama (2004: 322–23) reports hearing much the same.

1. Details in Ominde (1963, 127, 130). For more on Nyanza rainfall in general, see Morgan and Moss (1972). Currently the interagency Famine Early Warning Systems Network (FEWS NET), an outgrowth of the FEWS begun by the U.S. Agency for International Development in 1986, publishes regularly updated online news and maps of rainfall and vegetation, from satellite and ground observation, on Kenya and Africa as a whole.

2. Waligorski (1952) has found Luo farmers to be keen judges of soil types, and my impressions decades later support the finding.

3. About the nearest things to a general Luo ethnography published are Mboya (2001), Whisson (1964), Parkin 1978 (on Nairobi Luo), Cohen and Atieno Odhiambo (1989), and DuPré's annotated bibliography (1968). For the many aspects of daily life that upland Luo people share with near neighbors, despite major differences of origins and language between Nilotes and these Bantu-speaking people, one could do worse than to consult ethnographies on abutting Luhya (e.g., Wagner 1949, 1956; Abwunza 1997), and Gusii (LeVine and LeVine 1966, LeVine et al. 1996). Nancy Schwartz's work (esp. 1989) shows how much Luo like to conceive of themselves—even after their southerly migration has been slowed or stopped—as a people on the move.

4. Kenya is divided administratively into these units, listed in decreasing order of size: provinces, districts, divisions, locations, and sublocations. These correspond in some areas with ethnolinguistic groupings and subgroupings, but in others not. As results on ethnic breakdowns from the 1999 census were not yet available at the time of writing, those from the 1989 census are used here. The reader is cautioned that such figures, based largely on shifting self-definitions and always politically touchy, are subject to manipulation before being published.

5. Ruhlen (1991, 317–18). A student and devotee of the singularly influential language taxonomist Joseph Greenberg, Ruhlen acknowledges Lionel Bender's assistance on Nilo-Saharan tongues.

6. For studies of Luo origins and early migrations see Crazzolara (1950, 31–32); Ayany (1989), Ogot (1967, 28–41; 1999), Ochieng' (1975), Ehret (1976—on linguistic clues), Herring et al. (1984), Cohen and Atieno Odhiambo (1989). See also Evans-Pritchard (1965b, 209).

7. Onyango-Ogutu and Roscoe (1981, 133–39). For Acholi and Alur variants, see Crazzolara (1950, ch. 11; 1951, 225 and passim); Sudanese ones are compared in Lienhardt (1975).

Sometimes the stories of spear and bead occur separately, and sometimes with different personal names. Onyango-Ogutu and Roscoe use *juok*, discussed later, to convey the sacredness or mystical quality—in any case the specialness—of the particular spear. In this version of the story, present-day ethnonyms (Luo, Acholi, and Alur or "Aluru") are projected back in time. Many Nilotic groups appear to use similar stories to explain past divisions and migrations.

8. Hobley (1902, 26), Johnston (1904, 787), Roscoe (1915, 290), Hay (1972, 89–99).

9. Cohen and Atieno Odhiambo (1989, 12–15), on Gangu in Siaya.

10. On the sharpening of lineages that can occur under higher population densities, and their dissolution in declining population, see Shipton (1984). See also Glazier (1985) on Embu of central Kenya.

11. In the census of 286 Kanyamkago contiguous homesteads I organized in 1981–82, homesteads averaged 6.6 members. In Kenyan government surveys, the term *household* refers to the more precise "homestead" in rural contexts—that is, a polygynous group in a compound of houses is counted as one. Authors who have used "household" to refer to multihouse homesteads have sometimes used "hut" to mean house.

12. Patriliny means that descent and kin group membership are traced through males. Virilocality means that partners in marriage reside at the husband's natal home, in the Luo case usually among his patrilineal kin. In our 1981–82 Kanyamkago census, 37 percent of the male-headed homesteads were polygynous; 86 percent of the homesteads were classed as male-headed (n=270; sixteen further cases unclassifiable). But homestead headship is often debatable anyway. Much of what Abwunza (1997) writes on the informal power and influence of Logoli Luhya women, challenging patriarchy in times of economic hardship, holds true for the nearby Luo too; see also Francis (2000) on Luo themselves.

13. See also the writings of Whisson and the few of my own cited in the bibliography.

14. Luo landholding in colonial times is treated in Southall (1952), Wilson (1968), Pala (1977), Hay (1982), Shipton (1984, 1988, 1989), and Holmes (2000), among other available sources.

15. Segmentary lineages are ramifying kin groups, in this case based on real or putative patrilineal descent. While anglophone Luo conventionally use the term "clan" to refer to a patrilineal kin group of any depth, anthropologists usually prefer to reserve this term for such a group so deep in generations or so widely flung that its living members cannot trace all the links connecting them to its founding ancestor.

16. See Southall (1952), G. Wilson (1961), Parkin (1978), Goldenberg (1982), and Shipton (1989a) for more detailed discussions of Luo segmentary lineages.

17. See Fearn (1961, 84) on the plow's early introduction. The Kenya Ministry of Finance, Central Bureau of Statistics, *Rural Household Survey, Nyanza Province, 1970/1* (1977, 20); and Johnson (1980, 125) give figures on plow ownership and use in Nyanza Province.

18. Luo orthography in this study follows the standard form used by most Luo in writing; it is the one used in Stafford (1967), Clarke (n.d.), Blount and Padgug-Blount (n.d.) (this source further offers phonetic spellings), and Odaga (2004). Swahili terms are designated as such in the text.

19. Acland (1971) and Kokwaro (1979) give further general information on food crops of the region; Kokwaro (1972) has catalogued Luo names and medicinal and other uses of plants in Luoland.

20. For more on cotton growing in western Kenya see occasional reports in the *Empire Cotton Growing Review*, especially in the 1950s; Obara (1979).

21. Odede (1942), Southall (1952), Wilson (1968), and Ocholla-Ayayo (1979) describe the role of livestock in Luo culture. Ocholla-Ayayo explores the rich symbolism connecting particular livestock parts with particular kinship kin in marriages and other ritual meat distributions.

22. For statistics on livestock ownership, see Kenya Central Bureau of Statistics, *Rural Household Survey, Nyanza Province, 1970/1* (1977); DuPré (1968, 115-16). Figures on this topic are notoriously unreliable.

23. Cohen and Atieno Odhiambo (1989, 76-82) colorfully describe Luo animal marketing and part-time itinerant merchants engaged in it.

24. This preference for "upward" conversion is reminiscent of the Tiv in Nigeria studied by Bohannan (1955), with their "spheres of exchange." Barth (1967) prefers to imagine economic spheres among the Fur of Sudan as non-hierarchical (p. 157).

25. Compare Ferguson's (1990) and Comaroff and Comaroff's (1990) noteworthy expositions on parts of southern Africa, and Hutchinson's (1996) on the Nuer.

26. See Parkin (1978, 1980) for thorough discussions of Luo values surrounding bridewealth, polygyny, and the segmentary lineage model. A "domain of contestation" is where Ferguson locates cash and other liquid wealth in Lesotho families (1990).

27. Farmers in the most crowded areas where land is registered are beginning to "destock" and trying to optimize the ratio of productive to unproductive cattle in their herds. In South Nyanza this is done largely through exchanges with pastoralist Maasai. Having less need for draft animals, the Maasai sell bulls and oxen at prices advantageous to the Luo while paying high prices for Luo heifers and calves. But some Maasai have taken up cultivating.

28. Treated in the forthcoming volume in this series on credit across cultures.

29. *Loko* has many other meanings, including to change, alter, vary, transform, regulate, affect, renew, or transfer. In the reflexive *lokore* it can mean to repent. See Blount and Padgug-Blount (n.d., 85).

30. Glosses from Blount and Padgug-Blount (n.d.). Cf. Stafford (1967).

31. Analogous terms again appear in some Bantu languages—for instance, the Kinyarwanda term for barter places, *kabuzabwenge*. Pottier translates it as "'places that impede reasoning,' either [from the buyer's point of view] because frequent visits tend to result in total poverty or [from the seller's point of view] because all too frequent visits deplete one's food reserves" (Pottier 1986, 220).

32. Shipton (1990b) and sources referred to there discuss the process in other parts of Africa south of the Sahara.

33. See Fearn (1961), Kitching (1980), Hay (1982), Stichter (1982), and especially Francis (2000) for histories of Luo wage and off-farm work and a flavor of Luo occupational variety.

34. Lives of Luo urban workers, out-of-workers, and families are described and analyzed in Parkin (1969) and Grillo (1973) in Kampala, Parkin (1978) in Nairobi, and Buzzard (1982) in Kisumu, among other sources available.

35. See De Wilde (1967, 124 ff.). Working age is a problematic category, since boys and girls in western Kenya assist in and practice farming tasks of many kinds from the age of three or four.

36. UNILO (1972, 48), Grillo (1973, 50), Central Bureau of Statistics, *Rural Household Survey, Nyanza Province,* 1970/71, p. 36; cf. Moock (1976, passim), Francis (2000, 115–18). All figures on the topic must be treated with caution. Even if accurate, they may obscure local variation.

37. Population growth in Nairobi, for instance, appears at just under 5 percent per year between the 1969 and 1999 censuses (1999 census, 1:xxx).

38. Hay (1972), Kitching (1980), Goldenberg (1982), and Berman and Lonsdale (2002) are among the best places to learn about class differentiation, as it aligns and intersects with other sorts, in western Kenya. Obama's travel memoir (2004, ch. 15) offers a personal taste.

39. Francis (2000) on Luo and Abwunza (1997) on Logoli Luhya both vividly describe the gender tensions arising from disappointed hopes in migrant male labor and remittances. Along other lines, Goldenberg (1982) finds many subtle denials of class rifts in Luo country.

40. I have been struck, for instance, by the contrast between Kenya's relatively distinct ethnic territorial blocks and the much more thorough ethnic intermixture found in The Gambia.

41. This and the other population figures on ethnic group sizes come from the Kenya 1989 population census, the last so far to distinguish them.

42. The 1989 national census listed, in Nyanza Province, 2,517 "Kenyan Asian," 1,063 "Indians," 668 "Kenyan Arabs," 826 "Somalis" (many of these were restaurateurs), 71 "Pakistanis," 275 "other Asians," 544 "British," and 401 "other Europeans."

43. See Sytek's 1972 compilation of mutual interethnic stereotypes recorded among Luo and neighboring Gusii, Maasai, Nandi, and Kipsigis.

44. Christopher Ehret too has made this crucial point (1976, esp. 16–17), using linguistic evidence; nor was he the first. In this and other sections herein, my use of a "sliding" set of multiple criteria to define overlapping categories, knowing that no single category might contain them all, follows a tradition Wittgenstein identified as recognition of "family resemblances."

45. For history of Kenyan and Luo churches, see Barrett et al. (1973), and for updates, see others' occasional contributions to the *Journal of Religion in Africa.*

46. In matters concerning life's sequencing, a central concept is *chira,* referring both to the breaking of the strictest Luo cultural prohibitions about order and separations and to a metaphysically caused or facilitated wasting disease that results. Chira is discussed in Mboya (2001) and Parkin (1978).

47. The surveys and their methods and results are discussed in more detail in the forthcoming volume in this series on credit between cultures.

48. Alas, anthropologists who remain in their "field" sites for decades seem seldom to end up writing much ethnography, perhaps because of intellectual gridlock: every statement they attempt brings to mind exceptions.

49. Similarly, the Swahili verb *kuazima* means either to borrow *or* to lend.

50. In the Luo tongue a new transitive verb *uso,* to sell, from the Swahili *kuuza,* has become common and is tending to replace *ng'iewo* in that meaning. The use of a single term for both borrowing and lending, or one for both buying and selling, is fairly common in African languages. Both also occur elsewhere—for example, among the Aymara of Bolivia (Harris 1989, 248). Ludwig Wittgenstein posed this philosophical query: "Someone divides mankind

into buyers and sellers and forgets that buyers are sellers too. If I remind him of this is his grammar changed??" (Wittgenstein 1984, 18).

51. "Credit" in all its senses does not appear automatically translatable. While *holro*, for "loan," is listed in some dictionaries, it is absent from some (e.g., Stafford's 1967 Luo-English dictionary). One often hears the English noun *loan* used in DhoLuo sentences, particularly in reference to credit from afar.

52. Available Luo-English dictionaries offer no translation of *gowi* other than "debt." Compare *chulo* with the Nuer *col* (or *chol*) — which Evans-Pritchard translates "to indemnify" or "make recompense" (1956, 66, 228–29).

53. On borrower-lender misunderstandings about "interest" and about usury, see Shipton (1991) and the forthcoming volume in this series on credit between cultures.

Chapter 4: Entrustment Incarnate

1. Children's nursing of younger children, described in Ominde's brief study (1952) on Luo and in the unpublished notes of Paula Hirsch Foster on Acholi in northern Uganda at about the same time, still continues strong. It is no less as important among Bantu-speaking peoples nearby. See especially LeVine et al. (1996) on Gusii; and Weisner et al. (1997) on Luhya, including his rural-urban study of Kisa between 1968 and 1983 and a cross-cultural summary of African and worldwide findings. This chapter owes much of its insight to these sources. For an influential study on fostering in parts of Ghana, in some parts of which (for instance, western Gonja) more than a quarter of children are fostered at a time, see Goody (1982).

2. Nor are the systems of human and animal entrustment kept wholly discrete in practice, since many children help look after younger children at the same time as they are helping look after sheep, goats, chickens, or caged rabbits.

3. The system of classificatory kinship equating siblings and cousins by the same terms also reminds one of incest prohibitions. Luo expect the latter to apply ideally to any traceable agnatic or cognatic kin. Hutchinson (1996), comparing her findings on Nuer in the 1980s to Evans-Pritchard's in the 1930s, finds the range of prohibited kin to have narrowed, at least in practice. Polynesia, like some parts of Africa, is known for both classificatory (lumping) kinship terms for collaterals and for frequent child fostering.

4. The early training in sharing and passing along also struck Paula Hirsch Foster in Acholi country in the mid-1950s (dissertation typescript, n.d., pp. 63–64). She also remarked how theft of fruit from kin was deemed a moral failing, whereas theft from members of a different lineage or from Europeans was not. Compare the later discussion herein of cattle theft and raiding. Trampling of growing crops was not a matter of moral failing or sin but something not done ("penen"), not seen (pp. 78–79).

5. Blount and Padgug-Blount have translated *misumba* from Luo as "dependent, slave, worker, bachelor" (n.d., 102); Odaga as "bachelor, slave" (2004, 207). "No translation" might be as accurate. See Wilson (1968, 55–56, 124–25) on *misumba* ("a servant or a foundling brought up as a foster child") and *kimirwa* ("a child who was born before cattle were paid for his mother by her legal husband"—that is, what some would call a bastard): persons whose positions are comparable in some ways—and both stigmatized—but not necessarily identi-

cal, for instance in rights of inheritance. The point about fosterage and indebtedness holds for other Nilotic speakers. Okot p'Bitek's interpretations of two Acholi proverbs are telling. *Okwateng oloko ngeye ki polo* he translates as "a kite turns his back to the sky (his home)" with the explanation "Said of ungrateful person who turns his back on those who were kind to him, e.g. an orphan deserting a foster father in need." *Rwako yat itera* he translates, "He drives a long peg into the anus," explaining, "This is the manner in which night-dancers found dancing naked around houses were executed. A cruel death for a heinous crime. It is used as a protest against extreme ungratefulness e.g. an orphan beating up or ill-treating his foster mother" (p'Bitek 1985, 20, 14). P'Bitek also notes numerous proverbs that could be taken as warnings against foster mothers or stepmothers (*"Nyek meni pe meni—*Your stepmother is not your mother") (p. 11, see also 10), while noting, *"Ngat mumito nyac pa lakware myero omadi—*He who invites his grandson suffering from yaws must treat him,"—that is, cannot send him back to his parents just because of the expense (p. 11).

6. This component of respect for elders as a way of earning others' trust is a point insisted upon by Paula Hirsch Foster on the basis of her observations among Acholi in northern Uganda in the 1950s. (See Foster n.d., 93–94).

7. Beth Blue Swadener's and collaborators' book on western and central Kenya, *Does the Village Still Raise the Child?* well suggests, right in its chapter headings, how many parts of the story there are, though they appear here only in selected slices: In Maasai and Samburu country it takes grandmothers and a clan; on Kericho tea estates, child care centers and older siblings; on Kiambu coffee and tea estates, a weighing station and a supportive manager; in Embu district, tradition and intergenerational support; in Machakos district, preschool teachers as health workers; in Nairobi, money and partners; in Kisumu, *ayah*s (nursemaids) and preschools. Certainly it takes at least several sorts of nurturers, overseers, and guides in each such place, as varied forms of child entrustment form their weaves of particular color blends. Another volume might be written on how many sorts of children anyone in the region feels responsible to, or on how many elders and (by then) recent dead and ancestors the child grown up owes.

8. Hoddinott (1992) analyzes, in a rare econometric-ethnographic mix, the subtle bargaining in which potentially "rotten kids" interact with "manipulative parents" in Karateng, a Luo sublocation in western Kenya. He finds that (1) children do indeed provide important support for elders; (2) elders with more children get more support; and (3) parents use "heritable assets" (mainly, withholding land as future inheritance) to ensure that their grown children remember to send or bring them food and money—the threat of disinheritance being most credible where there are multiple offspring tacitly to compete. His findings broadly coincide with my observations and other ethnography from the region (see esp. Heald 1998).

9. For further details and discussion, see Parkin's detailed study of Kaloleni neighborhood, in Eastlands, Nairobi, in 1968–69 (1978), and his related earlier study of a ward of Kampala, Uganda (1969). On elite views on needier kin, see Obama 2004: ch. 15.

10. Sixty-nine percent of 393 adult lodgers (248 male, 145 female) were classed as "kin and affines of the [household] head" (1978, 89). This was the highest proportion among the several ethnic groups in Parkin's sample, which included Luhya, Gikuyu, Kamba, and others.

11. Buzzard (1982, 44). Sally Falk Moore (1986, 86) and Amy Stambach (2000, 106–7) discuss the "lending" of Chagga children between households in Tanzania, noting that where

lending out a girl benefits the lending family more than the borrowing one and thus incurs a reverse debt, this can sometimes be cleared by the lending family's using a portion of bridewealth received to her to repay the foster family. Here, as in Luo country, girls' dependence on sometimes exploitative hosts in fosterage can make them vulnerable to dangerous sexual liaisons.

12. Beth Pratt's doctoral research on Maasai in Monduli Juu, Tanzania, has made clear to me a considerable incidence of corporal punishment used in schools for infractions including absence — which can only discourage further attendance (Pratt 2003).

13. Since state-hired teachers frequently get sent far from home, these romantic liaisons and assaults must have an as yet unstudied effect of mixing gene pools.

14. The history and growth of the Islamic presence among Luo people and others in western Kenya — arising in part from the presence of Arab and southern Asian immigrant communities in the towns, and in part from circular migration to and sojourning on the Kenyan and East African coast — is the topic of a work in progress by Cynthia Hoehler-Fatton. I owe my own exposure to East and especially West African Qur'anic schooling mainly to Polly Steele, who has allowed me to participate at times in her own research.

15. None of this is to deny that Qur'anic tutelage, just like secular schooling, can sometimes turn exploitative, or physically abusive where corporal punishment is involved. The former risk is most evident (in parts of Islamic West and East Africa familiar to me) where disciples work in the fields or houses of their masters (also a training in itself, of course), or in a phase of training where disciples may be sent out begging to learn humility, to recompense or profit their masters, or all of these.

16. See Weisner et al. (1997), especially p. 21, and other sources they cite.

17. The topic of friendship among the African urban poor and destitute has up to now been covered best by works of fiction — for instance on Kenya, in novels by Marjorie Oludhe Macgoye (for instance, *Street Life*) and Meja Mwangi (*Going Down River Road*). In works like these, the importance of interethnic friendships is striking. For other parts of Africa, the Manchester and Rhodes-Livingstone schools of ethnography, focusing on southern Africa, provide models, particularly in the work of Clyde Mitchell, Philip Mayer, and others inspired by them, including Jeremy Boissevain.

18. Bankers learn this the hard way, in Luoland, when they try collecting cash loans in livestock form: an act for which Luo use the term *peyo*, to raid.

19. The most influential study of stock associateship is Gulliver's *The Family Herds*, on Jie and Turkana in Uganda and Kenya (1955). On Maasai and other pastoral peoples, see Waller (1988). Among the Kuria, to the southeast of the southern Luo along the Kenya-Tanzania border, Meinhard (n.d., 28) generalized that when borrowed cattle were stolen by some third party, the borrower was expected to try to recover them but bore no liability for them. On long-term livestock loans and clientage among the densely settled agrarian Chagga of Kilimanjaro, see Moore (1986, 67–70).

20. Asymmetry between figures for lenders and borrowers may be due partly to the small sample and the sensitive nature of the information.

21. In 1976, in Steven Johnson's sample of forty Luo-Suba homesteads surveyed in Wasweta-I Sublocation, near Migori, five homesteads had claims on twenty-three head of cattle lent to others in nearby sublocations; and ten held thirty-eight head borrowed. Four

homestead heads gave five cattle in 1975–76 to kin for the latter's own bridewealth payments, no other reason being given for gifts. Johnson found the homesteads to possess about six head on average, the range being from none (37.5 percent of sampled cases) to nearly eighty (Johnson 1980, 176–78).

22. Odede remarked (1942, 132) that although Luo commonly sent out their livestock to friends and relatives, they usually sent them where they could visit frequently to check up.

23. Elsewhere in Africa, livestock holders sometimes lend animals around to dodge tax assessment. For this and other reasons, herders all over East Africa have been observed to try to conceal how many animals they own. On funeral or post-funeral celebrations, however, Luo sometimes marshal herds of lineage or clan mates for stampedes to the borderlands of neighboring groups. See ch. 8.

24. They have been doing so for decades at least.

25. Expectations about who keeps the offspring of lent or entrusted animals vary by culture and circumstance. Some local and international aid agencies have lately attempted sensitively to base drought-recovery programs on loans of locally purchased livestock to herders who have lost stock, following local lending patterns in dividing up offspring. Oxfam's and other agencies' projects among Turkana of northern Kenya and WoDaabe of Niger are noted experiments, but I have been informed of similar, locally organized schemes in Zaire.

26. D. R. Crampton, "Laws of Secession for Chiefs and Headmen," in the *Indigenous Tribes and Their Customs* collection, book 4 (KNA, Syracuse Film 2081, reel 2). Thirty-three years later, Odede (1942, 131) similarly described a Luo "custom" whereby farmer A borrowed goats, assuring farmer B that he would return him a heifer until she calved. When the calf had weaned, it remained the property of farmer B.

27. These few remarks pertain more to ideals than to real behavior, which, in livestock exchange, is harder to discern.

28. When asked why, one informant claimed it was because one who repays a borrowed bull with a cow might later demand the cow back after it had given birth, claiming that the offspring itself was the repayment.

29. Of course, the borrower can exchange more or less any kind of livestock for any other before repaying. As noted earlier, Luo value male cattle about as much as female, partly because of the males' usefulness in plowing. This matters less to nearby Maasai herders, who tend not to plow: they prefer females to males because they produce milk and bear calves.

30. This does not imply, however, that the relations between fathers and sons, or mothers and daughters, are necessarily casual and friendly. Indeed, in this and many similarly patrilineal societies in Africa, father-son ties are usually much more formal than ties between mother's brother and sister's son, or than ties between members of alternate generations, which are often warm and easy-going.

31. Whether the conversion would be acceptable in reverse (that is, whether one who has borrowed a bullock could admissibly return four rams in their stead; or for a heifer, four ewes) is not clear to me, but my impression is that the preferred conversion is the "upward" one.

32. Schwartz (1989, 29–31) notes that animal species in western Kenya are associated with different ethnic groups and religions. Luo often associate goat meat with Bantu eating habits, thinking mutton and lamb more suitable for their own or for Christian food; but Luo have long used goats in ritual sacrifices.

33. See Shipton (1989, 37–38 and passim) on perception of "bitter money" attaching to the sale of roosters.

34. For more on livestock theft and raiding, see Hay (1972) on Luo, Sytek (1972) on Luo and neighbors, Heald (1998) on Gisu, and Waller (1998) on Maasai.

Chapter 5: Teaming Up

Epigraph. Okot p'Bitek, who recorded this Acholi proverb, offers this explanation of settling scores: "A borrows money from B and fails to pay it. After some time B borrows money from A's son or father, and refuses to pay it back" (p'Bitek 1985, 15). But this kind of shared accountability between generations need not refer just to money.

1. My information on so-called informal credit and debt in the following pages comes mainly from Kanyamkago Location, though much of it also applies in other Luo localities visited.

2. Arguably, of course, no single factor of production constrains farming by itself, but only factors in combinations.

3. Ominde's pioneering study of Luo girls' labor training (1952) remains useful, but the spread of primary schooling has changed the picture.

4. For varied but anthropologically informed discussions of rural labor exchange without cash payments, see Long (1985).

5. See Robertson (1987, 13, 43–46). His book compares Ghanaian, Sudanese, Lesothoan, and Senegambian cases of share contracts in one of the first sympathetic general treatments for Africa (see also Swindell 1985). In agriculture, Robertson notes, sharecropping beyond the household spreads risks of production but allows flexibility enough to accommodate uncontrollable shocks of weather, market failure, or policy change (p. 1). My discussion of loan and counter-loan partnerships owes much to Robertson's work.

6. Recall Gluckman (1965, esp. ch. 8) on Barotse obligations from property owning as socially cohesive. To him, laws of persons, things, and obligations all intersect (p. 271).

7. Fearn (1961, 226) recounts an anecdote about an African district council officer in Kisumu who tried to get his land weeded for money and received no help. When he changed his strategy and invited kin, killing a ram and providing beer, the task was finished immediately. See also Leach (1982, ch. 5), and on the amoral associations of cash outside Europe, see Parry and Bloch (1989).

8. Calculations of Kongstad and Monsted (1980, 73–74) based on 1974–75 Integrated Rural Survey data indicate that in nine districts of western Kenya, hired labor accounted for between 10 and 40 percent of farm labor in smallholder areas, but that Nyanza Province used less than other provinces, and South Nyanza District less than other districts except Bungoma. Whether sisters or female kin, or kin of different sexes, hire each other for cash as often as brothers or other male kin awaits further systematic study, but I suspect they do not.

9. Draft animals usually refer to oxen or other strong and manageable cattle, but in their absence some Luo also use donkeys, sometimes teaming them with cattle.

10. On the loan arrangements, my information on the Luo resembles Steven Johnson's on the nearby Luo-Suba on a number of points (1980, 122–29). See note 12.

11. An alternate explanation is that these exchanges are a concealed form of patronage by the owners of draft animals.

12. Steven Johnson (1979a, 1980) studied what he called "micro-level dependency relationships" among Luo-speaking Suba in southern Nyanza in the mid-1970s. For every day a "driver" spent with a borrowed plow team on his own field, he would spend two to three on the owner's (1980, 127 and passim). Known kin, however, borrowed without immediate or specific return favors. Plow team lending also partook of official patronage: Johnson found that "position-holders [in churches, schools, or government] who own plough teams loan them out freely far more often than do plough team owners who are not position-holders" (1980, 313).

13. Elizabeth Francis first brought to my attention the term *bar-wa-bar* for such share contracting in South Nyakach Location, Kisumu District. See her description of other ways of getting by there (2000).

14. Some farmers today charge their full brothers monetary rent for plow teams and plows. Cash seems to have made deeper inroads into kinship among the Isukha Luhya than among the Luo. The difference may reflect the heavier dependence of the crowded rural Isukha people on off-farm earning, a result of rural crowding.

15. My thinking on this topic since making my observations on the Kenyan side of the border has been enriched by Daivi Rodima's more recent research and forthcoming Brandeis University dissertation on Kuria on the Tanzanian side.

16. Rika and saga seem to be two merging categories in places. Odinga (1967, 13–14) uses the term *saga* to refer to past "communal cultivation" in "communities of anything from two to five hundred people" working "in strict rota" under the direction of the elders — perhaps a somewhat idealized view of the Luo past.

17. Compare, for instance, the *risaga* of the Gusii (Mayer 1951, 8–15), the *kibaga* of the Gisu of the Mt. Elgon area (Middleton 1965, 17), and the *lisanga* among the Isukha Luhya among whom I have stayed — all of whose names appear cognates of the Luo saga. *Rika* probably relates to the *morik* of the Kipsigis Kalenjin "southern Nilotic" speakers. Distinctions between rika and saga blur among the Luo. Steven Johnson (1980, 75) noted that Luo-Suba near Migori used the terms *rika, erika,* or *erikarene* for age sets of men circumcised together or in adjacent cohorts.

18. The comparison is from research of my own in the southwestern Colombian and Ecuadorian Andes, in 1973, and from a variety of sources on the Peruvian Andes. *Minga, minka,* and similar terms are also used in northern Andean languages other than Quechua.

19. Though see Parry and Bloch (1989) and Shipton (1989) for alternative arguments on the ties between cash and individualism.

20. *Shirika* (Sw., probably derives from Arabic and is a cognate of *rika*) varies in pronunciation in western Kenya (*sirika, shilika, silika*) among Nilotes and Bantu-speakers, some of the consonants being almost interchangeable. See Nakabayashi (1981) on new church-based shirika groups among Isukha Luhya. Kipsigis and other neighboring groups have comparable associations that sometimes go by other names.

21. The proverb on beer, recorded by Okot p'Bitek (1985, 20) as "Otwong kongo wille," comes from Acholi. He interprets it to mean "A gift or favour must be repaid." It recalls Marcel Mauss on the gift. Elizabeth Francis (2000, 118–22) offers more details on the smaller-

scale food lending and remittances in Kisumu District. Hers is a picture of rural women disappointed in the abilities or willingness of their urban migrant menfolk to provide. Of course, as she is aware, men's feeling obliged to keep up with reciprocities in beer drinks and bars is part of why women at home sometimes get the short end of the stick. See also Abwunza (1997) on nearby Maragoli Luhya.

22. A subtler kind of food sharing, as noted earlier, is child fostering.

23. Cf. Wagner (1956, 2:161) on southern Luhya.

24. The interviews were held individually and, for consensus, together. I have found that literate Luo and other Kenyans with a lifetime of dealing in extended kin ties have little trouble learning genealogical diagramming.

25. Note that the hypothetical transfers refer here to a male lender or giver (since males are likely to have to approve these major transfers of land, plow teams, or maize in bulk between homesteads). Gifts and loans by women tend to be smaller but more frequent. Whether the lenders or givers be men or women, many of the decisions are negotiated, or at least discussed, between spouses before they are acted upon. Real transfers between homesteads are not just transfers between individuals but also between groups. The informants here were men, but women's corresponding assumptions, on which my information is less complete, may not necessarily agree. I am also aware, of course, that a questioner's own assumptions can easily be projected onto respondents, and I have tried to separate these out.

26. Whether female lenders conversely favor female consanguinity remains to be answered.

27. Sixty-three percent mentioned friends, neighbors, or relatives in answer to the first question; fewer than 3 percent mentioned them in the third (Marco Surveys 1965, 53). Only 2 percent mentioned local shopkeepers, merchants, or traders in answering any question.

28. We repeated the Marco questions to small non-random samples of informants in Kagan and East Isukha in 1981–82. No answers were prompted. The main difference was that in our survey, different government offices were mentioned. For more on "informal" farm production credit in western Kenya, see Kenya CBS Rural Household Survey, Nyanza Province, 1970/1 (1975): 61–62.

29. These and other self-help groups, a subject of intense interest among foreign lenders seeking channels for lending and savings in Kenya as elsewhere, are discussed in the forthcoming volume on credit between cultures.

30. Even primary schooling posed financial problems for poorer families in requiring uniforms (shorts, shirts) and supplies. Primary students at the government-run Kagogo primary school responded to receiving pencils and exercise books as a treat.

31. Essays in Guyer (1987) discuss the shifting flows of rural-urban remittances in several African settings. Rural-urban migrations—for example, between Luo family homesteads and tea or sugar plantations—can also involve two-way transfers between kin.

32. E. P. Oranga, Assistant Agricultural Officer, Central Nyanza District, to Senior Agricultural Officer, Nyanza Province, June 28, 1949. KNA 3/1794 (Adm. 7/5/1), Syr. Reel 256, Sec. 10, p. 57. See also original typescript of the *Evidence* of the Kenya Land Commission, Syr. Reel 1925, sub-reel 10, on Central and South Kavirondo Districts.

Chapter 6: Marriage on the Installment Plan

Epigraph. *"Gen lubo lim":* "Trust follows bridewealth." Proverb recorded by Paula Hirsch Foster among Acholi in northern Uganda in the mid-1950s. Paula Hirsch Foster Archive, Boston University. Her translation. Also translatable as "hope follows (the) animals (or property)."

1. Use of the term *bridewealth,* referring to marriage payments from the bridegroom or his natal kin to the bride's natal kin, is sometimes criticized. For present purposes I use "marriage payments" or "marriage dues" to mean the same. As argued later, all these terms can place undue, or unduly exclusive, emphasis on economic or quantitative dimensions. East African Anglophones often use "dowry" to refer to what anthropologists call bridewealth, but since in other societies dowry can refer to goods given to a bride by her own parents (common in Asia), I prefer not to use it here to avoid confusion.

2. That is, when anthropologist Maurice Glickman wrote of bridewealth debts, among Nuer, as "the linchpin of the social organization" (1971, 313), he could as well have been speaking for their cousins the Luo.

3. Wilson (1968, 134) (no Luo translation); Goethe (1961, bk. 1, ch.9).

4. These forms of marriage recall what Claude Lévi-Strauss (1969) has described as direct and indirect exchange systems, respectively.

5. Lest any woman might go unmarried, Luo oral and now written tradition contains many warnings and exhortations. In one story the "ugly old hag," as soon as a brave fisherman Nyamgondho steps forward to marry her, rewards him with "vast numbers of cattle, sheep, goats, ducks, and barndoor-fowl" out from the lake (Onyango-Ogutu and Roscoe 1981, 140). (Note she does not turn into a princess: Luo do not go in for monarchy.) But Nyamgondho loses all the animals when he abuses her: they go back to the depths of the lake. Compare the Mumbo movement in chapter 9, on sacrifice, in which vast herds are promised to come from the lake to reward those who believe and sacrifice their large animals.

6. *Kend* refers to (a) marriage or a marriage ceremony process; in *kendruok,* marrying, the gerundive suffix *-ruok* can denote the married condition. *Kend* may be related to *kendo,* fireplace, or to *kende,* alone (as in made one).

7. See Kuper (1982) for a comparative summary of literature on bridewealth in southern Africa in "preindustrial" times and places. Kuper follows Henri Junod in using economic language of credit, debt, and reciprocity (see esp. pp. 27–29, 40, 110–12, 118, 120), but he also notes the rich symbolic meanings of cattle. Kuper finds that Zulu, Tswana, and other southern Bantu groups share concepts of cattle that associate them with white, semen, water, ash, rain, ancestors, snakes, and crocodiles: all "cool" things deemed to cause health and fertility. By contrast, red, menstrual blood, fire, lightning, and witches and their familiars he finds associated with "hot," causing sickness and sterility (pp. 19–20). Among Nilotes, too, one can construct such Lévi-Straussian schemata, but some feminists (both male and female) object to the implications about sex attributes or gender roles these can convey.

8. Parkin (1980, 219) uses this fact to contrast Luo to Giriama, Chonyi, and Digo, in eastern Kenya, who he has found conceptually and terminologically distinguish uxorial from genetricial rights. That Luo do not, in his view, befits the more unambiguous transfer of her roles as wife and mother (and rights over her children) to her husband's group and the greater

difficulty of divorce. Where payments are made mainly in cash instead of livestock, Parkin finds that separation and divorce are more likely.

9. Cf. Parkin (1978, 110; 1980) on monetization of bridewealth. He has seen clearer evidence of this than I have.

10. Potash's survey covered a population of 1,216, including 203 married men, some polygynous.

11. In 1990, a minimum expected was Ksh. 500 shillings (about U.S. $21), or, if she had reached Form Four in school and thus was deemed employable, Ksh. 1,000 ($42). But if the groom was a district officer or a secondary schoolteacher, known to have a salary of his own, he might be demanded Ksh. 2,000 ($84) and probably more cows too.

12. For more on Luo marriage and marriage payments, see Northcote (1907), Evans-Pritchard (1965b, ch. 12); Wilson (1968); Cotran (1968, ch. 16); Parkin (1978, 1980); Ocholla-Ayayo (1979); Cohen and Atieno-Odhiambo (1989, ch. 5). See also S. L. Johnson (1980, 166) on Luo-Suba; Sangree (1966, 14–15) on Tiriki; Comaroff (1980) on marriage payments and their meanings comparatively. Shadle (2000) adds history on Gusii.

13. On delayed bridewealth elsewhere in Kenya, see Håkansson (1988, 1989, 1990). His comparative analyses of delays found in different cultures are most interesting, though I query his classification of the Luo (1990, 162). See also Glazier (1985, 123) on the Mbeere of Embu district, Sangree (1966, 14–15) on the Tiriki of Kakamega District; and Glickman's library-based discussion of delayed bridewealth payments among Nuer (1971, 313).

14. Potash (1978, 392–93). In the fifty-one marriages of under five years' duration, forty-two (82 percent) had bridewealth payments reported still incomplete. In the thirty-five marriages five to ten years old, twenty-three (66 percent) had bridewealth incomplete. In 154 marriages over ten years old, eleven (7 percent) were incomplete, and of unknown duration, nine (8 percent). These figures exclude the twenty-five (7 percent) of the total of 347 cases where there was no bridewealth or it had been returned, and the thirteen (4 percent) of bridewealth status unknown. Potash's article argues (much like Buzzard's thesis on Kisumu city four years later) that Luo marriages are relatively quite stable (80 percent remain intact) despite tensions and hardships for women in them, and that reasons for a low divorce rate include child custody rules and land tenure conventions favoring men, and social stigma and practical difficulties that beset divorced women.

15. See Ocholla-Ayayo (1979, 182–83) on Luo, Hutchinson (1996) on Nuer, and Kuper (1982, 14–18) on southern Africa, for further discussions of bridewealth direction and hierarchy.

16. Ocholla-Ayayo (1979, 180–85). He bases his information on his own studies, on B. A. Ogot's oral historical texts, and on other sources. See also Crazzolara (1950, 1954) for further oral histories.

17. *Meko* is used in two ways: broadly, to refer to proper marriage process, and more narrowly to refer to the first major ceremonial sequence of marriage, which includes the bridal abduction and defloration. *Keny*, bridewealth, would seem to be a cognate of *kendo* (v.t., v.i.), to marry (man as subject, woman as object). The root of *moguedhi* (which blesses you) is *gueth* (or *gweth*), blessing. The transitive verb form, *guedho* (as *gwedho*), Grace Clarke (n.d.) defines as: "To protect, to bless, give an unsolicited gift, which is very acceptable" (p. G15).

18. Tension between "arranged" and self-selected "love" marriage, as acute in Luo coun-

try as anywhere, is a subject of Luo and Kenyan legends and written fiction, for instance Asenath Bole Odaga's short novel *The Villager's Son*. The theme also is a mainstay of Indian feature films Luo see in cities. (Is there anywhere young viewers do not favor autonomous love?)

19. Evans-Pritchard (1965a), Mboya (2001), Ocholla-Ayayo (1979), Wilson (1968). See also Cohen and Atieno Odhiambo's firsthand account (1989) and Odaga's (1971) novel about marriage negotiations and a quandary of elopement [*por*]).

20. I hope to offer elsewhere a more detailed treatment of custom, manipulation, and variation in Luo and western Kenyan marriage ritual, as part of a more general work in preparation about sequencing and seniority and their importance in "stateless" societies.

21. Interview with Emin Ochieng' Opere, January 1, 1992.

22. This distinction may be, as Wilson suspected, more important north of the Winam Gulf than south of it. I have heard people in Kanyamkago, south of the Gulf, use the terms *dhog keny* or *dho keny* for marriage cattle but deny recognizing the form *dho i keny*.

23. This assumes they live at her home, which of course they may not.

24. Grace Clarke's Luo-English dictionary identifies *dher powo* or *dher kede* as "earnest-cow; the cow chosen by the mother of the bride, generally a heifer which has not yet calved. This cow is seldom used in paying marriage dowry for sons' wives. It remains the mother's property. . . . But the cow's offspring is for dowries [bridewealth]" (p. D8).

25. Evans-Pritchard calls this *dher atum*, the cow of the bow (for shooting).

26. For instance, among southern Luo groups in Wilson's time, the left foreleg of a sacrificed goat, *bat chien*, the "portion of the ghost," stayed in the owner's *dala* "for his women" (Wilson 1968, 111). Evans-Pritchard wrote that the more northerly Kenya Luo evidently distinguished more sharply between named and unnamed cattle than southerners (1965b, 239).

27. Wilson wrote, "a [marital] union which is very successful continues the transference of gifts until the death of both parties . . . ideally" (1968, 116, para. 77). But the transference can go on beyond that.

28. According to Evans-Pritchard's research in Alego in the 1930s (1965b, 234), when a bride's father had called her back home, the groom had the right to send his brothers to a marketplace to drag her back to him. Surely a matter of perspective.

29. On delayed marriage payments and risk avoidance, cf. Glazier's similar findings (1985, 123–24) on the Mbeere of Embu District, or Glickman's Nuer reanalysis noted earlier (1971, 313–15). I think the practice of protracting bridewealth quite common in eastern and southern Africa.

30. Notes written by Dan Odhiambo Opon in response to an early draft of this chapter.

31. Wilson (1968, 115, para. 72). I have no evidence that this custom has changed.

32. Evans-Pritchard (1965, 237, 240–41), Wilson (1968, 115–18, esp. para. 75). Gordon Wilson in the mid-1950s recounted a "recent" case in central Nyanza where, of a total of twenty-two animals a groom had paid, none was left to repay him after his bride's kin deducted "those killed in his honor and those retained for virginity, children and neighbors" (1968, 115, para. 75).

33. Wilson (1968, 134, para. 168).

34. I have heard indications, however, that there may lately be a trend toward separating or divorcing women's claiming children for themselves and their natal families, in contravention of "customary law." Further study is to be desired.

35. *Wero* (n., v.i.), to reclaim marriage cattle, may be the root of both *weruok* (divorce, marital or family disagreement) and *werruok* (forgiveness, pardon), which some dictionaries treat under one spelling. The following information on return of bridewealth in cases of divorce comes from my interviews in Kanyamkago in 1990–91, supplemented with information from Wilson (1968, 139), Potash (1978), and other sources as noted.

36. For a different view, emphasizing a bride's detachment from her natal family and supposedly full incorporation into the groom's lineage, see Wilson (1968, 118, para. 85).

37. It can be a tough call, however, whether to call a couple "separated," since many Luo and other western Kenyan marriages involve spatial separation without necessarily emotional or economic estrangement.

38. Wilson (1968, 139, para. 194) specified in 1955 the expectation in "north and central" Luo tribes that for each *son* alive at the time of divorce, *three* heifers and a bull would be retained by the woman's father, whereas only *two* heifers and one bull for each *female*. "Southern tribes" counted one heifer less for each side. Potash's case material from 1973 to 1975 (1978, 389 and passim) indicates three cattle deducted for a son, two for a daughter; and Parkin likewise noted Luo divorce settlement deductions of "three cattle for a son and two for a daughter, or the monetary equivalent" (1980, 200).

39. Wilson and his informants found, in 1955, that "the natural increase of the bride wealth animals are not returnable should the marriage be dissolved. These are regarded as compensation for the loss of the daughter and for the services enjoyed by the husband for the period of the union" (1968, 140, para. 199. Cf. 133, para. 158).

40. Wilson (1968, 133, para. 159). Another section notes, though, that animals that have died in an epizootic must be replaced. Why the difference is not clear. But it may bear obliquely on crop insurance schemes tried and failed in western Kenya in recent decades.

41. Indeed, according to Wilson's 1955 report, the prior expenditure of animals by the bride's family in marriage ceremonies could wipe out all of their debts or even indebt the husband (Wilson 1968, 139, para. 194).

42. Discussed further, with case illustrations, in Potash (1978) and Buzzard (1982).

43. From his notes made in response to a rough draft of this work.

44. Wilson (and/or his informants) opined that in the past, elopers, usually people in reduced circumstances, usually at least eloped with the *intention* to carry out their customary obligations of ceremony and bridewealth when they could; but, he wrote, "Today, however, these unions take place without this sincerity" (1968, 122, para. 105). This assumption we may question.

45. E. P. Oranga, a locally hired Assistant Agricultural Officer, Central Nyanza, to Senior Agricultural Officer, Nyanza Province, June 28, 1949. KNA 3/1794 (ADM. 7/5/1). For references to other cases where humans have been pledged or pawned in African famines, see Shipton (1990b).

46. *Nyar* can mean daughter or pudenda (perhaps a synecdoche). Clarke (n.d., N32). *Osiep*, most often used to mean friend (or friendship), is also used to mean *fiancé* or betrothed (p. O25).

47. Wilson (1968, 121, para. 103). More is said about this in the following chapter and a separate work in progress about sequencing.

48. Presumably, if the father is dead by then, the debt falls upon the son, but this was not made explicit to me.

49. In the past, as recalled in Kanyamkago in the early 1980s, the bodies of pre-pubic girls have been buried outside homesteads, unlike most other bodies, which are buried within. In the 1980s it was considered proper to bury them outside the left side of the entrance (if any), as one faced outward. This is considered an outcast position. Such a burial is expected to anger the spirit of the dead, who may return with vengeance—all of which may serve as an incentive to keep young girls alive and healthy.

50. Cf. the Nuer case, where "bridewealth is not necessarily paid to the bride's father, but distributed among a number of homesteads in payment of debts that the bride's father contracted" (Glickman 1971, 314; see also Evans-Pritchard 1951).

51. That marriage's entrustments and obligations help keep society stitched together is, of course, a hard thing to prove, but one probably harder to disprove.

Chapter 7: Debts in Life and Death

Epigraph. Abwunza (1997, 32) on Logoli Luhya in western Kenya. Her point was not to disparage elders but to suggest a continuity between them and ancestors.

1. Schwartz (2000) describes living and dead women's influence among Luo and Luhya (Luyia). Igor Kopytoff distinguishes African from European understandings of ancestors, arguing, "Once we recognize that African 'ancestors' are above all elders and to be understood in terms of the same category as living elders [ambivalently, as both benevolent and punishing], we shall stop pursuing a multitude of problems of our own creation" (1997, 418). Separately, Ludwig Wittgenstein criticized philosophers who say, "'After death a timeless state will begin,' or: 'at death a timeless state begins,' and do not notice that they have used the words 'after' and 'at' and 'begins' in a temporal sense, and that temporality is embedded in their grammar" (1984, 22c).

2. A book-length study of Luo funerals has yet to be published, but the topic is worthy of one. See Mboya (2001, chs. 12, 13) for one Luo's respected if somewhat idealized account of funerary customs, originally in Luo. See also Wilson (1968, 130–32 and passim) and Goldenberg (1982, ch. 9).

3. Goldenberg (1982, 284), in a year of research in Karachuonyo, south of the Winam Gulf, in the mid-1970s, found no burial societies such as I found elsewhere in southern Nyanza a few years later.

4. Consider Wittgenstein's remark in 1949: "The concept of festivity. We connect it with merrymaking; in another age it may have been connected with fear and dread" (1984, 78e). In many minds horror and humor can, of course, connect in some of the same activities.

5. They may have objected because the ceremony pierced their territory and symbolically effeminized it and its inhabitants, evoked ancient battles, or to some Christians perhaps seemed primitive or superstitious. But the custom of tero buru is performed throughout Luo country.

6. Some of the animals slaughtered may in fact have been oxen; I did not see them all.

7. Evans-Pritchard (1965b, 241), Wilson (1968, 134). Evans-Pritchard writes it *"apongo*

dhuonga." According to Wilson, the *dho miluhini* and the "status and sacrificial animals" are paid in full this time round, but *dho i keny* not. I have inadequate information to generalize on how far young women are challenging the sororate now.

8. While childless or "barren" *wives* who died were replaced, according to Wilson, childless or "barren" *widows* who died were not. This difference seems to suggest that compensation was being made for the husband's own damaged prestige or feelings of disappointment.

9. For further discussion of Luo widow inheritance and "ghost marriage," see Evans-Pritchard (1965b, 239–40); on this and "ghost marriage," see Wilson (1968, 127–32) and Ocholla-Ayayo (1976, 144).

10. Sharp distinctions between *genitor* (biological father) and *pater* (social father) are common among Nilotes but may not be universal in East Africa. See Heald (1998, passim) on Bantu-speaking Gisu in southeastern Uganda.

11. From "Laws of Succession for Chiefs and Headmen," memorandum of 3 September 1909, in KNA, "Indigenous Tribes and Their Customs," Book 4, Syracuse Film 2081, Reel 2.

12. Anon., undated memorandum, "Nilotic Kavirondo (Luo)," in KNA, "Indigenous Tribes and Their Customs," Book 4, Syracuse Film 2081, Reel 2.

13. Among Shilluk people in Sudan too, relatives of Luo, an ostensible object of homicide compensations was to provide means for the bereaved group to obtain a new wife, and there the rates were commensurate with bridewealth rates (Howell 1952, 118).

14. "Ghost marriage" is also a well-known feature of Nuer kinship as described by Evans-Pritchard (1951, 109–13 and passim), and much has been written on it since his account was published. See, for example, Hutchinson (1996, 61–62, 175–76, 337, and passim*)*.

15. According to Gordon Wilson, murder in late colonial times was thought to bring *kwer,* a state of profound dishonor and supernatural impurity, on an individual—but not necessarily *chira,* a still more extreme form that can involve the family, lineage, or oganda as a whole (Wilson 1960, 182). Both kwer and chira call for ritual purification.

16. For more on Luo ghostly vengeance, see Evans-Pritchard (1950), Ocholla-Ayayo (1976, 180), Abe (1978). Evans-Pritchard, among Nuer, also noted an understanding of a metaphysical poisonous blood bond formed between killer and killed. It is reminiscent of Mauss's "spirit of the gift" that boomerangs, but here it is one inverted for harm.

17. Anon., report of District Commissioner's *baraza,* at Chief Chweya's Camp, Central Kavirondo, September 15, 1933. KNA, "Indigenous Tribes and Their Customs," Syracuse Film 2081, Reel 2. Peacemaking by ritually splitting an animal in half has also been practiced among Bantu speakers nearby, including Gisu in southeast Uganda (Heald 1998, 167–69).

18. Among Tiv in eastern Nigeria in the 1950s, for instance, a mother's brother could be fined for the misdeed of a sister's son resident with him (assaulting an accuser of theft), despite Tiv patriliny (Bohannan 1989, 72ff.; see also Gluckman 1965, 258–59). The blame had to be shared.

19. Heald's detailed ethnography of Gisu in southeast Uganda, just across the border from Luo country, doubts kin share much sense of group responsibility, despite their patrilineal ideology.

20. One might object that sharing responsibility also diffuses it (such that one need be less concerned about punishment if the punishment will fall on one's kin instead of oneself)

and that this gives an incentive to selfish or careless behavior. But shared responsibility might, for better or worse, just raise anxiety about moral behavior all around.

Chapter 8: In the Passing

1. Emin Ochieng' Opere provided important insights and impressions contained in this chapter, based on his experience in Sakwa, Kanyamkago, and elsewhere (including his own father's death and mother's status as a widow). Wilson (1968), also useful, focuses its discussion of inheritance almost entirely on land.

2. In the view of Suzette Heald (1998), writing on Gisu in nearby southeast Uganda, father-son tension over bridewealth and inheritance of land and animals, in conditions of crowding and shrinking resources, is a tap root of the anger and frustration that moves men to violence, even within their own generations. Something of the same applies to Luo.

3. I thank Nancy Schwartz for bringing this proverb to my attention. The plural suffix *gi* alludes to the woman's affines. Some would say the proverb demeans females by likening them to wild animals. Parental sex bias can set heirs at odds (see Obama 2004: 318-19).

4. Occasionally, however, a son does remove his animals from his father's herd without permission. Then people talk and worry for the family.

5. Killing or serious injury between brothers is also more maladaptive than between co-wives, since brothers share more genes. It would be unsurprising to find it rarer.

6. Wilson used the term *keyo nyinyo* but not *sikukuu*, noting that it happened about a year after the burial, and that *jokakwaro*, close agnates, take "all they can carry away, things such as spears, harps, bows, arrows, and today furniture, dishes, blankets, etc." (1968, 131–32). Goldenberg notes that in Karachuonyo in the 1970s, the division of a male's property was called *gilaro*, "they dispute the ownership" (1982, 291).

7. The wildness of these animals used for funeral garb contrasts with the domesticity of animals sacrificed.

8. On this point Emin Ochieng' Opere conducted and taped an interview for me at which I could not be present, with three Kanyamkago male elders, two Luo (one from Nyakach) and one Luo-Suba, in late 1991.

9. Sojourners preparing to leave Kenya for overseas—symbolically akin to dying—are sometimes struck to see their neighbors competing among themselves for sloughed household possessions, even simple ones, whether as mementos (a weak taste of bereavement?) or merely as much-needed goods.

Chapter 9: Blood, Fire, and Word

Epigraph. "Religioeser Glaube und Aberglaube sind ganz verschieden. Der eine entspringt aus Furcht und ist eine Art falscher Wissenschaft. Der andre ist ein Vertraun." Wittgenstein (1984, 72). By these definitions most humans are capable of both.

1. Evans-Pritchard, 1964 preface to Hubert and Mauss (1981, viii). See also Evans-Pritchard (1956, 286), where he writes of "religion . . . and its central act of sacrifice." Compare Meyer Fortes, "there is no ritual institution as central to all but a minority of the scriptural and the non-scriptural religions of mankind" as sacrifice (1980, v).

2. Except in the sense that (as many Muslims would say) there is nothing that is not a part of religion. Religion is a hard term to define, but any minimum definition ought to include both thought and action, and perhaps some reference to sacred or ultimate things, beings, powers, principles, or conditions, including ultimate causes and effects, or ultimate beginnings and ends.

3. Charles G. Seligman's early works on southern Sudan directed the attention of a generation of Anglophone anthropologists, including James Frazer and later Seligman's student Edward Evans-Pritchard, to the Shilluk case. Frazer's *The Golden Bough*, that multivolume work most often read in abridgements (e.g., 1924, 266–67, 294–95), in turn influenced so much of literature in English and other languages. Seligman's work on Shilluk and Evans-Pritchard's on Nuer (1956, 1965a) strongly influenced that of Evans-Pritchard's close colleague Godfrey Lienhardt on Dinka (1961). For a few of the many available discussions of sacrifice among Nilotic-speaking peoples in southern Sudan, see Beidelman (1974), Burton (1987), and Hutchinson (1996). See also Middleton (1960) on Sudanic-speaking Lugbara and Whyte (1997) on Bantu-speaking Nyole, both near Lwoo Nilotic-speaking peoples.

4. Some later scholars have doubted the accuracy of J. Roscoe's ethnography in other details. My own informant here said chicken but more likely meant rooster.

5. I thank E. S. Atieno Odhiambo, D. A. Masolo, and Benzburg Nyakwana for their kind help in clarifying the nuances between these terms. Grace Clarke, a missionary, translates *liswa* as "sacrifice" (and *golo liswa*, "offer a sacrifice"), and in another meaning as "sacrilege, wickedness (?)" (n.d., L11). *Dolo* carries the alternate meaning of "to bend" (Stafford 1967, 99) or "To wind up, curl, roll up, wind round" or "to make cicatrices round the navel" (Clarke n.d., D20–21), but any link between these meanings and sacrifice is unclear. A ritual specialist with the power to make sacrifices is called a *jadolo*. Clarke translates *misango* as "sacrifice, oblation" (p. M30); I imagine it comes from *sango*, "to assemble, as of a crowd of people; unite, join together as one," or "to eat copiously of various kinds of food" (p. S3). *Sango ni* means to join with someone in alliance, and *ring' sango* refers to specified portions of meat that a bride's kin take to her groom's. Sacrifice in Swahili is *sadaka*, which also translates as alms, or an act of charity inspired by religion.

6. Blood sacrifice refers in the present work to offerings involving, but not necessarily limited to, the movement of blood outside the body. G. A. S. Northcote wrote in 1907 that Luo make offerings (to the sun, he thought) "at all important occasions in their daily life" (p. 63). At least some Luo sacrifice may fit, without too much twisting, Evans-Pritchard's four-part model (to him, a standard sequence in the Nuer case) of presentation, invocation, consecration, and immolation (Evans-Pritchard 1956, 207–15; Hoehler-Fatton 1996, 125–28, 233n).

7. Ocholla-Ayayo (1979, 190–93); also Goldenberg (1982, 287n.); cf. Lienhardt (1961, 24), on Dinka.

8. Nuer and other Nilotic peoples have shown a penchant for choosing oxen (castrated bulls) for other major sacrifices. Beidelman's explanation (1974) is psychosexual, focusing on the ox's sex-neutrality as a symbol of tamed human sexuality and virility as a product of culture and society. Other explanations are possible, including the ox's relative docility as a safety consideration for the sacrificer, its high meat and fat mass, or its superfluity for reproducing a herd.

9. On the one and many for Nuer, see Evans-Pritchard (1956, 200), although it is hard to

disentangle Christian from indigenous Nuer influences in this ethnography. On Luo concepts of divinity see Stam (1910), Whisson (1962a), Hauge (1974, 97–105), Ocholla-Ayayo (1976, 166–170), Schwartz (1989); Hoehler-Fatton (1996, 233). Many Luo speak of *Ruoth* (chief, lord) Nyasaye. Nyasaye may come from *sayo*, to beseech, implore, or adore, and it is a Bantu term too, but its provenance is uncertain. Interestingly (according to Grace Clarke's dictionary), it has a plural, *nyiseche*, though I have seldom heard it. *Were*, as a name for divinity, is shared with Luhya (Bantu-speaking) peoples. On whether a supreme creator deity is an indigenous idea among Luo and other lake-river Nilotic people, see p'Bitek (1971, ch. 3, esp. pp. 50–51) for a naysayer's view on "central" (Uganda) Luo; for a yea-sayer's see Hauge 1974, 97–104). Many others have weighed in and no clear consensus has emerged. Some Bantu-speakers near Luo also associate sun with creator divinity—for instance, Kuria (Ruel 1997, 18–19). *Juok* (or *Jwok*), almost certainly a cognate of *Jok* in many Nilotic tongues, is versatile but not easily translatable. It has been used to mean a higher all-pervasive power, spirits or minor divinities, life force, certain diseases, "fetishes," anomalies, and many other things, including magic, witchcraft, and inexplicable things generally, not all referring to personified powers. See Ogot (1999, ch. 1) on Luo, and p'Bitek (1971) for discussions and comparisons. Customs of spirit possession associated with Jok or Juok may have diffused to Lwoo from Bantu speakers, particularly from Nyoro (Curley 1973, 152–55).

10. Ruel found comparable uses of chyme among Kuria in the 1950s (1997, 24) and cites too its use among Luhya, Gusii, Gisu, and other Bantu-speaking people. He calls "non-sacrificial ritual killing" the killing done to cleanse or "to restore, ward off misfortune or render healthy or whole" without being directed to any "personally conceived spirit, deity or being" (pp. 89, 94).

11. That is, the term *sacrifice* has been given many more meanings than this. Bourdillon (1980, 10) files them under three rubrics: ritual slaughter or surrender of an animal or person; the giving up or destruction of something for something else usually more valuable; and the (Christian) mass. Here we are concerned mainly with the first, but secondarily with the others too.

12. Several of the contributors in Bourdillon (1980), notably Bourdillon, Beattie, and Sykes, take comparable, "polythetic" approaches to sacrifice. So does Sperber, writing on Evans-Pritchard and the Nuer (1987, esp. 24–26). Polythetic is Needham's term (1975).

13. For a clear example, see Hauge (1974, 102): a prayer accompanying a sacrifice of a hen, thanking Nyasaye for protection from enemies and for a safe childbirth, and for protection from a flood, and at the same time requesting protection from elephants and a good setting of the sun ("may the sun set well"—in Luo idiom of prayer, this last connotes a general blessing).

14. Hutchinson (1996) discusses a case where a young Nuer woman appears to make a grab for authority by taking it upon herself to sacrifice a goat, a task usually performed by a senior man. Debates over such cases can refresh and perpetuate the rules they ostensibly break. Separately: on the importance of what he calls "liturgy," or sequences presumed timeless, see Rappaport (1999).

15. Cohen and Atieno Odhiambo relay that in the Luo district of Siaya, some elders report that defenseless widows were sometimes "seized and sacrificed" to protect a new boat for use on Lake Victoria (1989, 110n). Ruel notes that among Kuria of Butimbaru area, similarly and occasionally, in crisis time, a man was taken and ritually trampled upon until ex-

creta and stomach contents emerged from his anus, and these were taken and scattered upon arable fields. Also, among Kuria, death by drowning is widely thought to be followed by plentiful crops (Ruel 1997, 26). Ruel does not, however, consider these actions to have the same "personal, reciprocating force" as does killing a goat for giving an ancestor a "skin-garment" (p. 24n; recall Mauss's "force" in the gift that "compels a return"). Whether such acts and understandings spread from Nilotic Luo to adjacent Bantu-speaking Kuria, vice versa, reached both from some third source, or arose independently in both is unknown.

16. The Mumbo movement (along with a mid-to-late twentieth-century protest movement among Bukusu and other Luhya that was called Dini ya Msambwa) is discussed in most detail by Audrey Wipper's sociological history (1977), which, along with Maxon (1989), is a main source for this discussion.

17. Indeed, some historians would include early Christianity itself as a "revitalization" movement resembling movements in many parts of the world today among people at the edges of empires.

18. Ogot (1971) and Hoehler-Fatton (1996), the main sources on which I base this brief section on the history of the Roho movement, tell the story of Alfayo Odongo Mango and the Kager-Wanga in more detail; see also Holmes (2000). I have known a number of Roho church adherents in southern Nyanza. For purposes of this study, anyone who calls him- or herself a Christian is called a Christian.

19. Mango's exact birthplace, his marital history, and the details of his turn to Christianity are all in some dispute (see Hoehler-Fatton 1996, 18–21).

20. Two of the major ones, the Ruwe Holy Ghost Church of East Africa, and the Musanda Holy Ghost Church of East Africa, regard Alfayo Odongo Mango as their founder and savior, while the Holy Spirit Church of East Africa turns its reverence instead to Jacobo Buluku (or Jacob Muruku) (d. 1938).

21. Note the nuance between this and the Nuer *kuk*, implying immediate reciprocation.

22. *Iwewa erichowa, kaka wan wawejo matimonwa marach* (Forgive us our sins as we forgive those who have done evil to us). Roho practitioners also encode this into their own version of Luo as *Enjunjo Umpedhinjo, chocho njond njonjuwi sovesindnjo sompodh* (Hoehler-Fatton 1996, 144).

23. King James translation. In the *Book of Common Prayer* and some other scripture, "debts" is rendered as "trespasses," the common idea being the crossing into obligation.

Chapter 10: Conclusion

Epigraph. Douglas (1980, 3).

1. Note, too, the connotations of gender and seniority in the English *patronage*, from Latin *pater*, father.

2. Patrick Bateson (1988) cites dunnocks, hedge sparrows, in particular. Some birds unrelated genetically, but mated and cooperating in care of offspring, are known to adjust their behavior to each other's. When one mate dies, the other works harder at care of the offspring, even at the expense of chances to find other mates (Bateson 1988, 22). None of this implies that individuals necessarily do what is best for their species.

3. A subtler but broader lesson derivable from Luo and Kenyan tradition—in a way

more of a stretch, and one that goes against some demagogic pronouncements—is that all living creatures share something precious: that humans can find a great deal in common with other animals. This applies to the cattle with which (or with whom) Luo and other Nilotic men in particular like so closely to identify: so closely that they let them cement their friendships in extended loans and stand in for themselves and other humans in marriage dues and in occasional sacrifices. But humans also have something in common even with the creatures coming out of the bush to eat crops—as suggested again in the story of the lost spear and bead. Here human and elephant, despite past enmity between their kinds, prove able to overcome that—and their difference of species, sex, and age—to form a common understanding that benefits both.

Glossary

All terms defined are in the Luo language (DhoLuo) unless otherwise indicated.
Key: Ar. = Arabic; Eng. = English; Swa. = KiSwahili.

Note on Luo pronunciation: Letter *a* pronounced as in *lot*. Other vowels all have both tensed and relaxed (closed and open) pronunciations (for "harmony" between root vowels and suffix vowels). They are: *e* (as in *bait* or *get*), *i* (as in *see* or *sit*), *o* (as in *bone* or *law*), *u* (as in *who* or *could*). Consonants: *g* (as in *get*), *h* (as in *hat*), *r* (somewhat staccato); other single consonants as in English. Combined consonants: *ch* (as in *chalk*), *dh* (as in *the*), *ng'* (as in *thing*), *ny* (as in *Kenya*), *th* (as in *bath*).

ajuoga, pl. *ajuoge* or *ajuoke* (n.) a diviner, a magician, or an adept at medicines. See also *jwok*.

anyuola (n.) (from *nyuol*, n., v.i., [to give] birth) a lineage; a group constituted by putative descent from a common ancestor.

ayie (or *ayieth*) (n.) loop of rope for staking a cow; or, preliminary token of agreement for a marriage, sometimes in cash, given to prospective bride's father or parents.

baraza (from Swa.) (n.) chief's or other government official's public meeting.

bar-wa-bar or *bar wabar* (n.) term for reciprocal exchange labor, implying chopping or splitting (*baro*, v.t.), symmetry.

biero (n.) placenta (buried, signifies belonging).

buru (n.) literally, ashes. A mortuary ceremony (or set of ceremonies) conducted upon a delay after a first funeral (liel). (v. *tero buru*).

chamo (v.t.), *chiemo* (v.i.), to eat (food = *chiemo*); figuratively, to take over (as by extortion), to accept (as a gratuity or bribe), or to absorb. Object of verb may be a person or thing: to eat (of) someone.

chira (n.) (no direct translation) a mystical principle of harm ensuing from an important breach of custom (usu. of a prescribed sacred sequence or a rule of avoidance), or a wasting illness or other affliction resulting as a manifestation, sometimes as a shared or collective fate. (Cf. *kwer*.)

chiri (n.) long rainy season (usu. February to May). (Cf. *opon.*)

chon gilala (long ago), and *thu tinda* (perhaps from *tho tinda*, "sickness/death, spare me"): phrases used respectively to begin and end narrative (as myth, legend, or tale).

dala, pl. *mier*. (n.) homestead, may be enclosed and/or multi-house. Also sometimes translated village or hamlet. (Cf. *pacho.*) *Goyo dala:* to build a home.

dani (n.) elder woman or grandmother; a term of respect. (Cf. *pim.*)

debe (n.) (from Swa.) a tin of roughly standard size used for measuring foodstuffs (often four to six per sack).

dero, pl. *dere* (n.) granary hut.

dhiang', pl. *dhok* (n.) cattle. *Dhok keny, dho i keny*, marriage cattle (core component of bridewealth).

dho (-) as noun, mouth; as prefix: language (of), as in DhoLuo, the Luo tongue.

dhoot, pl. *dhoudi* (n.) literally, mouth (*dho*) of house (*ot*), i.e., doorway. Used figuratively to refer to people who descend from one house or woman. In this usage, a clan, or smaller or shallower agnatic descent group (i.e., patrilineage).

dilo (v.t.) to exorcise or placate a spirit or ghost. Cf. *jadilo.*

dini (n.) church, sect.

doho (n.) council or meeting, as of elders; court or tribunal. Also, polygyny.

dolo (v.t.) to sacrifice, to offer up. *Jadolo:* one with enough seniority, knowledge, or authority to sacrifice. Sometimes trans. as "priest."

ero kamano. "correct, it is there"—a thanking acknowledgment.

geno (v.t., v.i., also n.) to hope, trust, expect, depend on, or have confidence in.

gogo (n.) Fishing net or trap enclosure.

gowi, pl. *gope*. (n.) debt, obligation. *Bedo gi gowi*, literally, to sit (or remain) with a debt: to (continue to) be obligated to someone.

gunda bur, pl. *gundni bur* (n.) stockaded multi-house settlement, or remains of one, usu. from pre-European times.

gweng', pl. *gwenge* (n.) a land area of clan or other group. *Jaduong' gweng'*, leader of same.

guedho or *gwedho* (v.t.), to bless, protect. *Gueth* or *gweth* (n.) blessing.

harambee (exclam., n.) (Swa.), literally, "let's pull together." Used in Kenya (since time of President Jomo Kenyatta) for a gathering or occasion for self-help contributions.

holo (v.t.) to lend or borrow. *Holro* (n.): a loan; borrowing, lending.

ja-, pl. *jo-* (personal prefix): (the) one (who, of, or from . . .).

jabilo, pl. *jobilo* (n.) medicine man, diviner (of an *oganda*).

jachien, pl. *jochiende* (n.) vengeful or malevolent spirit (or ghost) of one deceased.

jadak, pl. *jodak* (n.) literally, stayer or dweller. Allochthon, land client; a relatively late arriver to an area or a descendant of such person(s); sometimes loosely and misleadingly translated alien, tenant, squatter. Term connotes inferiority to original settlers (cf. *wuon lowo*), and not belonging.

jachan (n., adj.) poor (person). Contrast *jamoko.*

jadil or *jadilo*, pl. *jodil, jodilo* (n.) an exorcist; one who placates or wards off ghosts or spirits.

jaduong, pl. *jodongo* (n.) literally, big or great one; used for an elder, usu. a male.

jagam, pl. *jogam* (n.) go-between or pathfinder (in arranging a marriage).

jamoko (n., adj.) rich, wealthy (person). Contrast *jachan.*

japidi, pl. *jopidi* (n.) child nurse or caregiver (for younger child).

japith, pl. *jopith* (n.) wealthy person.

jok (n.) unseen force (see also *jwok*).

jokakwaro (n.) (group of) people who descend (through males) from the same grandfather. Sometimes extended to refer to descendants of ancestor at a further remove.

jwok, juok (n.) Unseen force with many refractions (pl., *juogi*), also used to refer to witchcraft. *Jajuok,* evil magician or sorcerer, manipulator of *juok.*

kend (n.) marriage. *Kendo* (v.i., v.t.) to marry. *Kendruok* (n.) condition of marriedness.

keyo (n.) literally, branch; used for a lineage segment or faction. *Keyo nyinyo* (n.) a ceremony marking the end of a mortuary ritual.

kom, pl. *kombe* (n.) stool, usu. three- or four-legged.

kuon. pl. *kuonde* (n.) Luo staple food consisting of boiled grain meal (of sorghum, millet, or latterly maize). Comparable to polenta or firm hominy grits. Sometimes trans. "stiff porridge" or "Luo bread." (Swa.) *ugali.*

kwer (n.) hoe. Also, a prohibition, its breach, or a condition of dishonor and mystical impurity resulting (cf. *chira*).

lemo (n., v.i.) prayer, invocation, or solemn statement. *Pogo lemo* (n.), dividing prayer (testament for inheritance) or gathering for a partitioning.

libamba, pl. *libembni* (n.) A large and genealogically deep patrilineal descent group.

liel, pl. *liete* (n.) grave or funeral.

liswa (n.) a kind of sacrificial ritual, as with a propitiatory offering. (cf. *misango*)

lowo, pl. *lope* (n.) land, soil, or area; may imply attachment to a particular person or group (see *wuon lowo* and *jadak*).

matatu (n.) (from Swa.) taxi van or minibus, often a pickup truck with rear enclosure.

meko (v.i.), *mako* (v.t.), to marry according to traditional sequence. (cf. *kendo.*)

mbuta (n.) Nile perch (Lates niloticus), a species of large, edible freshwater fish known for predation.

miaha (n.) new or trial bride.

mikayi (n.) senior co-wife (cf. *nyachira, reru*)

mihambi (n.) thing pledged as security (as against a loan).

misango, pl. *misengni* (n.) a kind of sacrificial ritual. (Cf. *liswa.*)

ngege (n.) tilapia (Oreochromis, esp. O. esculentus), freshwater fish within cichlid family, a favorite Luo food.

ng'iewo (v.t.) to purchase; sometimes also to sell (as in *ng'iewore,* to sell oneself). To engage in a sale.

nyachira (n.) second co-wife (cf. *mikayi, reru*)

nyar (n.) daughter (of). *Nyar ot,* house daughter (trial affine, often connoting subservience).

Nyasaye (pl. nyiseche, less often heard). Name of supreme creator divinity, sometimes associated with the sun. Oft. trans. "God."

nyieka, pl. *nyieke* (n.) co-wife. *Nyiego:* envy or jealousy.

nyombo (v.i.), *nywomo* (v.t.) to transfer animals (in marriage payment).

oganda, pl. *ogendni* (n.) Untranslatable; an indigenous socio-political unit or location consisting of a federation of clans, sometimes deemed a chiefdom or quasi-chiefdom, sometimes trans. "tribe." See also *ruoth.*

omena (n.) A minnow (Rastrineobola argentea), a freshwater sardine.

ongeza (from Swa.) (n.) something extra; amount of something voluntarily added by a seller upon completing a sale, for goodwill.

opon (n.) short rainy season (usu. October to December). (Cf. *chiri.*)

osiep (n.) friend, or one betrothed.

osuri (n.) house roof spire, associated with a senior male's dominion.

ot, pl. *odi* or *udi* (n.) house. Also sometimes trans. "hut."

pacho, pl. *mier*. (n.) homestead, or home. Sometimes trans. "village," "hamlet." Cf. *dala*, a term with overlapping meanings.

pesa (n.) money. *Pesa makech*, bitter (i.e., morally tainted and dangerous) money.

peyo (v.t.) to raid or plunder; also latterly used for official confiscation, as in a foreclosed mortgage.

pim, pl. *pimbe* (n.) old woman, esp. one who instructs children, as by storytelling.

piny, pl. *pinje* (n.) country, home area, nation.

piro (v.t.) to dun, demand payment of a debt. Syn. *bandho*.

por (n.) elopement (i.e., marriage without ceremony or wealth transfers)

puro (v.i., v.t.) to farm, cultivate. *Japur:* farmer.

remb (n.) blood, life force, binding substance.

reru or *rero* (n.) third or more junior co-wife (cf. *mikayi, nyachira*)

riembo (v.t.) Literally, to drive (away); used in reference to animal(s) entrusted by one to another's care at a distance.

rika (n.) rotating self-help labor association. Cf. *shirika*.

ring'o (n.) meat, flesh.

riso (n., v.t.) a ceremony considered to bind a marriage.

RoSCA (Eng. acronym) rotating saving and credit association (i.e., a contribution club with turn taking).

rundo (v.t.) to buy, sell, turn over, or confuse.

ruoth or *rwoth*, pl. *ruodhi*. (n.) No exact translation. Indigenous leader of an oganda, sometimes translated as chief or ruler. Also used to mean "Lord," as in Ruoth Nyasaye.

saga (n.) large, festive labor gathering, sometimes involving work in competition.

serkal (from Ar. through Swa. serikali) (n.) government. Sometimes used loosely with reference to any or all officialdom, incl. that of large corporations.

shirika (from Swa.) a collective work group, often church based.

simba (n.) (probably from Swa. for "lion") in a homestead, a bachelors' house.

singo (n.), pledge, security, collateral, guarantee, obligation; (v.t.) to pledge, to put up as a guarantee, to barter for something else, to owe. *Singruok* (n.) contract, covenant, security, promise, testimony (see also *mihambi*).

siwindhe (n.) in a homestead, nursery for children or dormitory for girls, some perhaps from other homesteads.

ter (n.) process of widow "inheritance" or takeover. *Jater,* one who takes the widow.

thur (n.) homeland, home locality.

tim, pl. *timbe* (n.) custom.

tipo (n.) ghost, spirit, shade, apparition of deceased person.

ugali (n.) (Swa.) boiled grain meal (*unga*). See *kuon*.

ulaya (n.) (Swa.) alien country, land of outsiders.

waro (v.t.) to redeem. Cf. *wero*.

wat (n.) kinship. *Owadwa,* our kinsperson.

wero (v.t.), to take back (as with cattle upon a divorce).

wuon or *won*, pl. *weg* (n.) father, master; sometimes trans. "owner." *Wuon dala,* homestead head or most senior resident. *Wuon lowo:* master of the land, or original settler or descendant thereof; autochthon, one who belongs.

yie, pl. *yiedhi* (n.) canoe, boat.

Bibliography

Abe, Toshiharu. 1978. "A Preliminary Report on *Jachien* among the Luo of South Nyanza." Discussion Paper 92, Institute of African Studies, University of Nairobi.

Abila, Richard O. 2003. "Fish Trade and Food Security: Are They Reconcilable in Lake Victoria?" In Food and Agriculture Organization (U.N.), *Report of the Expert Consultation on International Fish Trade and Food Security*, pp. 128–54. F.A.O. Fisheries Report 708. Rome: F.A.O.

Abrahams, R. G. 1998. *Vigilant Citizens: Vigilantism and the State*. Cambridge, U.K.: Polity Press.

Abwunza, Judith. 1997. *Women's Voices, Women's Power: Dialogues of Resistance from East Africa*. Peterborough, Ontario: Broadview.

Acland, Julien Dyke. 1971. *East African Crops*. Hong Kong: Longman.

Appadurai, Arjun, ed. 1986. *The Social Life of Things: Commodities in Cultural Perspective*. Cambridge: Cambridge University Press.

Ardener, Shirley. 1964. "The Comparative Study of Rotating Credit Associations." *Journal of the Royal Anthropological Institute of Great Britain and Ireland* 4(2):201–29.

Ardener, Shirley, and Sandra Burman. 1996. *Money-Go-Rounds: The Importance of Rotating Savings and Credit Associations for Women*. Oxford, U.K., and Washington, D.C.: Berg.

Ayany, Samuel G. 1989 (1952). *Kar Chakruok mar Luo* (On the Origins of the Luo). Kisumu: Lake Publishers.

Baier, Annette. 1995. "Trust and Its Vulnerabilities." In Annette Baier, ed., *Moral Prejudices*, pp. 131–51. Cambridge, Mass.: Harvard University Press.

Banfield, Edward C., assisted by Laura Fasano Banfield. 1958. *The Moral Basis of a Backward Society*. Glencoe, Ill.: Free Press.

Banholzer, P., and J. K. Giffen. 1905. *The Anglo-Egyptian Sudan*, ed. Count Gleichen.

Baran, Paul. 1957. *The Political Economy of Growth*. New York: Monthly Review Press.

Barclay, Albert H. 1977. *The Mumias Sugar Project: A Study of Rural Development in Western Kenya*. Ph.D. thesis, Columbia University.

Barrett, David D., George K. Mambo, Janice McLaughlin, and Malcolm McVeigh, eds. 1973. *Kenya Churches Handbook*. Kisumu: Evangel Publishing House.

Barth, Fredrik. 1967. "Economic Spheres in Darfur." In Raymond Firth, ed., *Themes in Economic Anthropology*, pp. 149–89. London: Tavistock.

———. 1969. *Ethnic Groups and Boundaries: The Social Organization of Culture Difference*. London: Allen and Unwin.

Bateson, Patrick. 1988. "The Biological Evolution of Cooperation and Trust." In Diego Gambetta, ed., *Trust: The Making and Breaking of Cooperative Relations*, pp. 14–30. Oxford: Basil Blackwell.

Beattie, John H. M. "On Understanding Sacrifice." In M. F. C. Bourdillon and Meyer Fortes, eds., *Sacrifice*, pp. 29–44. New York and London: Academic Press.

Beidelman, T. O. 1966. "The Ox and Nuer Sacrifice: Some Freudian Hypotheses about Nuer Symbolism." *Man* (new series) 1(4):453–67. Reprinted 1974 in Robert A. LeVine, ed., *Culture and Personality*. Chicago: Aldine.

Benedict, Ruth. 1989. *The Chrysanthemum and the Sword: Patterns of Japanese Culture*. Boston: Houghton Mifflin. Orig. pub. 1946.

Berman, Bruce, and John Lonsdale. 2002. *Unhappy Valley: Clan, Class and State in Colonial Kenya*. Columbus: Ohio University Press.

Berry, Sara. 1984. "The Food Crisis and Agrarian Change in Africa: A Review Essay." *African Studies Review* 27(2):59–112.

Besteman, Catherine, and Lee V. Cassanelli, eds. 1996. *The Struggle for Land in Southern Somalia: The War Behind the War*. Boulder, Colo.: Westview.

Biebuyck, Daniel. 1963. *African Agrarian Systems*. London: Oxford University Press.

Blount, Ben G., and Elise Padgug-Blount. n.d. *Luo-English Dictionary*. Nairobi Institute of African Studies, University of Nairobi. Occasional Publication.

Boas, Franz. 1969. *The Ethnography of Franz Boas*, ed. H. Codere. Chicago: University of Chicago Press.

Bohannan, Paul. 1955. "Some Principles of Exchange and Investment among the Tiv." *American Anthropologist* 57:60–70.

———. 1959. "The Impact of Money on an African Subsistence Economy." *Journal of Economic History* 19(4):491–503. Reprinted in George Dalton, ed. 1967. *Tribal and Peasant Economies*. New York: Natural History Press, pp. 123–35.

———. 1989. *Justice and Judgement among the Tiv*. Prospect Heights, Ill.: Waveland. Orig. pub. 1957.

Bourdieu, Pierre. 1977. *Outline of a Theory of Practice* (*Esquisse d'une théorie de la pratique*). Cambridge: Cambridge University Press. Orig. pub. 1972.

Bourdillon, Michael F. C. 1980. Introduction to M. F. C. Bourdillon and Meyer Fortes, *Sacrifice*, pp. 1–28. London and New York: Academic Press.

Branstrator, D. K., J. T. Lehman, and L. M. Ndawula. "Zooplankton Dynamics in Lake Victoria." In Thomas C. Johnson and Eric O. Odada, eds., *The Limnology, Climatology and Paleoclimatology of the East African Lakes*, pp. 337–56. Amsterdam: Gordon and Breach.

British East Africa Protectorate, *Annual Reports*. London.

Burton, John. 1987. *A Nilotic World: The Atuot-Speaking Peoples of the Southern Sudan*. Westport, Conn.: Greenwood Press.

Butterman, Judith M. 1979. *Luo Social Formations in Change: Karachuonyo and Kanyamkago, c. 1800–1945*. Ph.D. thesis, Syracuse University.

Buzzard, Shirley. 1982. *Women's Status and Wage Employment in Kisumu, Kenya*. Ph.D. thesis, American University, Washington, D.C.

Carlebach, Julius. 1980. "Juvenile Prostitution in Nairobi." In Erasto Muga, ed., *Studies in Prostitution: East, West, and South Africa, Zaire and Nevada*, pp. 70–129. Nairobi: Kenya Literature Bureau.

CGIAR (Consultative Group for International Agricultural Research), International Potato Center, 2006. Kenya land map, http://www.cipotato.org/DIVA/data/DataServer.htm, May 31, 2006.

Chernoff, John M. 2003. *Hustling Is Not Stealing: Stories of an African Bar Girl*. Chicago: University of Chicago Press.

Clarke, Grace A. n.d. *Luo-English Dictionary*. Kendu Bay, Kenya: East African Publishing House.

Cohen, A. S., L. Kaufman, and R. Ogutu-Ohwayo. 1996. "Anthropogenic Threats, Impacts and Conservation Strategies in the Africa Great Lakes: A Review." In Thomas C. Johnson and Eric O. Odada, eds., *The Limnology, Climatology and Paleoclimatology of the East African Lakes*, pp. 575–624. Amsterdam: Gordon and Breach.

Cohen, Abner. 1969. *Custom and Politics in Urban Africa: A Study of Hausa Migrants in Yoruba Towns*. Berkeley: University of California Press.

Cohen, David William. 1983. "Food Production and Food Exchange in the Precolonial Lakes Plateau Region." In Robert I. Rotberg, ed., *Imperialism, Colonialism, and Hunger: East and Central Africa*, pp. 1–18. Lexington, Mass.: D. C. Heath.

Cohen, David William, and E. S. Atieno Odhiambo. 1989. *Siaya: The Historical Anthropology of an African Landscape*. London: James Currey; Nairobi: Heinemann Kenya; Athens, Ohio: Ohio University Press.

———. 2004. *The Risks of Knowledge: Investigations into the Death of the Hon. Minister John Robert Ouko in Kenya, 1990*. Athens: Ohio University Press.

Comaroff, Jean, and John L. Comaroff. 1990. "Goodly Beasts, Beastly Goods: Cattle and Commodities in a South African Context." *American Ethnologist* 17(2):195–216.

Comaroff, John, ed. 1980. *The Meaning of Marriage Payments*. New York: Academic Press.

Cotran, Eugene. 1968. *Kenya*, Vol. 1: *The Law of Marriage and Divorce. Restatement of African Law* series, 1, ed. Anthony N. Allott. London: Sweet and Maxwell.

Crazzolara, J. P. 1950, 1951, 1954. *The Lwoo*, Part I: *Lwoo Migrations*. Part II: *Luo Traditions*. Part III: *Clans*. Verona: Museum Combonianum.

Crump, Thomas. 1981. *The Phenomenon of Money*. London, Boston, and Henley: Routledge and Kegan Paul.

Curley, Richard T. 1973. *Elders, Shades, and Women: Ceremonial Change in Lango, Uganda*. Berkeley: University of California Press.

Deacon, Terrence. 1997. *The Symbolic Species: The Co-evolution of Language and the Human Brain*. London and New York: Allen Lane (Penguin).

Deutsch, M., D. Canavan, and J. Rubin. 1971. "The Effects of Size of Conflict and Sex of Experimenter on Interpersonal Bargaining." *Journal of Experimental Social Psychology* 7:258–67.

De Wilde, John C. 1967. *Experiences with Agricultural Development in Tropical Africa*. 2 vols. Baltimore, Md.: Johns Hopkins University Press.

De Wolf, Jan. 1977. *Differentiation and Integration in Western Kenya: A Study of Religious Innovation and Social Change among the Bukusu*. The Hague: Mouton.

Dietler, Michael, and Ingrid Herbich. 1998. "*Habitus*, Techniques, Style: An Integrated Approach to the Social Understanding of Material Culture and Boundaries." In Miriam T. Stark, ed., *The Anthropology of Social Boundaries*, pp. 232–63. Washington and London: Smithsonian Institution Press.

Douglas, Mary. 1966. *Purity and Danger*. Washington, D.C.: Praeger.

———. 1980. *Edward Evans-Pritchard*. New York: Viking.

Douglas, Mary, and Baron Isherwood. 1978. *The World of Goods*. Harmondsworth, U.K.: Penguin.

Downs, Richard E., S. P. Reyna. 1988. *Land and Society in Contemporary Africa*. Hanover, N.H.: University Press of New England.

Du Bois, Cora. 1969. *People of Alor: A Social-Psychological Study of an East Indian Island*. Cambridge, Mass.: Harvard University Press. Orig. pub. 1944.

DuPré, Carole E. 1968. *The Luo of Kenya: An Annotated Bibliography*. Washington, D.C.: Institute for Cross-Cultural Research.

Ehret, Christopher. 1976. Aspects of Social and Economic Change in Western Kenya. In B. A. Ogot, ed., *Kenya Before 1900*, pp. 1–20. Nairobi: East African Publishing House.

Evans-Pritchard, Edward E. 1940. *The Nuer*. Oxford: Oxford University Press.

———. 1946. "Nuer Bridewealth." *Africa* 16(4):247–57.

———. 1950. "Ghostly Vengeance among the Luo of Kenya." *Man* 50(133):86–87.

———. 1951. *Kinship and Marriage among the Nuer*. Oxford: Oxford University Press. Reprinted 1990.

———. 1956. *Nuer Religion*. Oxford: Oxford University Press. Reprinted 1974.

———. 1965a. "Luo Tribes and Clans." In Edward E. Evans-Pritchard, *The Position of Women in Primitive Societies and Other Essays in Social Anthropology*. London: New York: Free Press, 205–27. Previously published in 1949 in Rhodes-Livingstone Journal 7: 24–40.

———. 1965b. "Marriage Customs of the Luo of Kenya." In Edward Evans-Pritchard, *The Position of Women in Primitive Societies and Other Essays in Social Anthropology*, pp. 228–44. New York: Free Press. Orig. pub. 1950 in *Africa: Journal of the International African Institute* 20(2):132–42.

Falola, Toyin, and Paul Lovejoy, eds. 1994. *Pawnship in Africa: Debt Bondage in Historical Perspective*. Boulder, Colo.: Westview Press.

Fearn, Hugh. 1961. *An African Economy: A Study of the Economic Development of the Nyanza Province of Kenya, 1903–1953*. London: Oxford University Press.

Ferguson, James. 1990. *The Anti-Politics Machine: "Development," Depoliticization, and Bureaucratic Power in Lesotho.* Cambridge: Cambridge University Press.

———. 1992. "The Cultural Topography of Wealth." *American Anthropologist* 94(1):55–73.

Firth, Raymond. 1963. "Offering and Sacrifice: Problems of Organization." *Journal of the Royal Anthropological Institute* 93(1):12–24.

———. 1964. Introduction to Raymond Firth and Brian Yamey, eds., *Capital, Saving and Credit in Peasant Societies.* London: George Allen and Unwin.

Firth, Raymond, and Brian Yamey, eds. 1964. *Capital, Saving and Credit in Peasant Societies.* London: George Allen and Unwin.

Fleisher, Michael L. 2000. *Kuria Cattle Raiders: Violence and Vigilantism on the Tanzania/Kenya Frontier.* Ann Arbor: University of Michigan Press.

Flynn, Karen Coen. 2005. *Food, Culture, and Survival in an African City.* New York: Palgrave Macmillan.

Fortes, Meyer. 1980. Preface to M. F. C. Bourdillon and Meyer Fortes, *Sacrifice,* pp. i–xix. New York and London: Academic Press.

Foster, Paula Hirsch. n.d. Unpublished notes on the Acholi female life course. Mugar Memorial Library, Boston University.

Foucault, Michel. 1970. *The Order of Things: An Archaeology of the Human Sciences.* New York: Pantheon.

———. 1978. *The History of Sexuality.* vol. 1. Trans. Robert Hurley. New York: Pantheon.

Francis, Elizabeth. 2000. *Making a Living: Changing Livelihoods in Rural Africa.* London and New York: Routledge.

Frank, Andre Gunder. 1967. *Capitalism and Underdevelopment in Latin America: Historical Studies of Chile and Brazil.* New York: Monthly Review Press.

Frazer, James George. 1924. *The Golden Bough: A Study in Magic and Religion* (abridged ed.). London: Macmillan.

Freedman, J. L., and S. Fraser. 1966. "Compliance without Pressure: The Foot in the Door Technique." *Journal of Personality and Social Psychology* 4:195–202.

Fukuyama, Francis. 1995. *Trust.* New York: Free Press.

Gambetta, Diego. 1988. "Can We Trust Trust?" In Diego Gambetta, ed., *Trust: The Making and Breaking of Cooperative Relations,* pp. 158–75. Oxford: Basil Blackwell.

Geertz, Clifford. 1962. "The Rotating Credit Association: A 'Middle Rung' in Development." *Economic Development and Cultural Change* 10(3):241–63.

Geissler, P. Wenzel, Erdmute Alber, and Susan Reynolds Whyte, eds. 2004. *Grandparents and Grandchildren. Africa* (Journal of the International African Institute) 74(1).

George, Susan, and Fabrizio Sabelli. 1994. *Faith and Credit: The World Bank's Secular Empire.* Boulder, Colo.: Westview Press.

Girard, René. 1977. *Violence and the Sacred* (*La Violence et le sacré*). Baltimore, Md.: Johns Hopkins University Press. Orig. pub. 1972.

Glazier, Jack. 1985. *Land and the Uses of Tradition among the Mbeere of Kenya.* Lanham, Md.: University Press of America.

Glickman, Maurice. 1971. "Kinship and Credit among the Nuer." *Africa* (Journal of the International African Institute) 41(4):306–19.

————. 1974. "Patrilinity among the Gusii and Luo of Kenya." *American Anthropologist* 76:312–18.

Gluckman, Max. 1965. *The Ideas in Barotse Jurisprudence.* New Haven, Conn.: Yale University Press.

Godelier, Maurice. 1977. *Perspectives in Marxist Anthropology.* Trans. Robert Brain. Cambridge, London, New York, Melbourne: Cambridge University Press. Orig. pub. 1973.

Goethe, Johann Wolfgang von. 1961. *Elective Affinities,* trans. R. J. Hollingdale. Harmondsworth, U.K.: Penguin. Orig. pub. 1808.

Goldenberg, David. 1982. *We Are All Brothers: The Suppression of Consciousness of Socio-Economic Differentiation in a Kenya Luo Lineage.* Ph.D. thesis, Brown University. Ann Arbor, Mich.: University Microfilms.

Goldsworthy, David. 1982. *Tom Mboya: The Man Kenya Wanted to Forget.* Nairobi and London: Heinemann; New York: Africana Publishers.

Good, David. 1988. "Individuals, Interpersonal Relations, and Trust." In Diego Gambetta, ed., *Trust: The Making and Breaking of Cooperative Relations,* pp. 31–48. Oxford: Basil Blackwell.

Goody, Esther. 1970. "Kinship Fostering in Gonja: Deprivation or Advantage?" In Philip Mayer, ed., *Socialization: The Approach from Social Anthropology,* pp. 51–74. London: Tavistock.

————. 1982. *Parenthood and Social Reproduction: Fostering and Occupational Roles in West Africa.* Cambridge: Cambridge University Press.

Goody, Jack, and S. J. Tambiah. 1973. *Bridewealth and Dowry.* Cambridge: Cambridge University Press.

Gregory, Chris. 1982. *Gifts and Commodities.* New York: Academic Press.

Grillo, Ralph. 1973. *African Railwaymen: Solidarity and Opposition in an East African Labour Force.* Cambridge: Cambridge University Press.

Gudeman, Stephen. 1986. *Economics as Culture: Models and Metaphors of Livelihood.* London: Routledge and Kegan Paul.

Gulliver, P. H. 1955. *The Family Herds: A Study of Two Pastoral Tribes in East Africa, the Jie and Turkana.* London: Routledge and Kegan Paul.

Guyer, Jane. 1981. "Household and Community in African Studies." *African Studies Review* 24(2/3):87–138.

————, ed. 1987. *Feeding African Cities.* Manchester, U.K.: Manchester University Press, for International African Institute.

————, ed. 1995. *Money Matters: Instability, Values and Instability in the Modern History of West African Communities.* Portsmouth, N.H.: Heinemann.

————. 2004. *Marginal Gains: Monetary Transactions in Atlantic Africa.* Chicago and London: University of Chicago Press.

Guyer, Jane, and Pauline Peters. 1987. "Conceptualizing the Household." *Development and Change* 18(2):197–213.

Håkansson, Thomas. 1988. *Bridewealth, Women and Land: Social Change among the Gusii of Kenya.* Stockholm: Almqvist and Wiksell.

———. 1989. "Family Structure, Bridewealth, and Environment in Eastern Africa: A Comparative Study of House-Property Systems." *Ethnology* 28(7):117–34.

———. 1990. "Descent, Bridewealth, and Terms of Alliance in East African Societies." *Research in Economic Anthropology* 12:149–73.

Harris, Olivia. 1989. "The Earth and the State: The Sources and Meanings of Money in Potosi, Bolivia." In Parry, Jonathan, and Maurice Bloch, eds., *Money and the Morality of Exchange,* pp. 232–68. Cambridge: Cambridge University Press.

Hart, Keith. 1982. *The Political Economy of West African Agriculture.* Cambridge: Cambridge University Press.

———. 1988. "Kinship, Contract, and Trust: The Economic Organization of Migrants in an African City Slum." In Diego Gambetta, ed., *Trust: The Making and Breaking of Cooperative Relations,* pp. 176–93. Oxford: Basil Blackwell.

Hauge, Hans-egil. 1974. *Luo Religion and Folklore.* Oslo: Universitetsforlaget.

Haugerud, Angelique. 1984. *Household Dynamics and Rural Political Economy among Embu Farmers in the Kenya Highlands.* Ph.D. thesis, Northwestern University. Ann Arbor, Mich.: University Microfilms.

———. 1995. *The Culture of Politics in Modern Kenya.* Cambridge: Cambridge University Press.

Hay, Margaret Jean. 1972. *Economic Change in Luoland: Kowe 1890–1945.* Ph.D. thesis, University of Wisconsin. Ann Arbor, Mich.: University Microfilms.

———. 1982. "Women as Owners, Occupants and Managers of Property in Colonial Western Kenya." In M. J. Hay and Marcia Wright, eds., *African Women and the Law: Historical Perspectives,* pp. 110–23. Boston: Boston University African Studies Center.

Hayley, T. T. S. 1970. *The Anatomy of Lango Religion and Groups.* Cambridge: Cambridge University Press. Reprinted: New York: Negro Universities Press (of Greenwood Press). Orig. pub. 1947.

Heald, Suzette. 1998. *Controlling Anger: The Anthropology of Gisu Violence.* Oxford, Kampala, and Athens, Ohio: James Currey, Fountain Publishers, and Ohio University Press. Orig. pub. 1989.

———. 2000. "Tolerating the Intolerable: Cattle Raiding among the Kuria of Kenya." In G. Aijmer and J. Abbink, eds., *The Meanings of Violence,* pp. 101–121. Oxford: Berg.

Herbich, Ingrid, and Michael Dietler. 1993. "Space, Time and Symbolic Structure in the Luo Homestead: An Ethnoarchaeological Study of 'Settlement Biography' in Africa." In Juraj Pavúk, ed., *Actes du XIIe Congrès International des Sciences Préhistoriques et Protohistoriques,* pp. 26–32. Bratislava.

Herring, R. S., D. W. Cohen, and B. A. Ogot. 1984. "The Construction of Dominance: The Strategies of Selected Luo Groups in Uganda and Kenya." In Ahmed Ida Salim, ed., *State Formation in Eastern Africa,* pp. 126–61. Nairobi, London, and Ibadan: Heinemann Educational Books.

Hobley, C. W. 1902. "Nilotic Tribes of Kavirondo." In *Eastern Uganda: An Ethnological Survey,* pp. 26–35. Royal Anthropological Institute Occasional Papers, no. 1.

———. 1903. "Anthropological Studies in Kavirondo and Nandi." *Journal of the Anthropological Institute of Great Britain and Ireland* 33:325–59.

Hoddinott, John. 1992. "Rotten Kids or Manipulative Parents: Are Children Old Age Security in Western Kenya?" *Economic Development and Cultural Change* 40(3):545–65.

Hoehler-Fatton, Cynthia. 1996. *Women of Fire and Spirit: History, Faith, and Gender in Roho Religion in Western Kenya.* Oxford: Oxford University Press.

Holmes, Jane Teresa. 2000. *A Home for the Kager: Negotiating Tribal Identities in Colonial Kenya.* Ph.D. thesis, University of Virginia.

Howell, P. P. 1952. "Observations on the Shilluk of the Upper Nile: The Laws of Homicide and the Legal Functions of the *Reth.*" *Africa* 22(2):97–119.

Hubert, Henri, and Marcel Mauss. 1981. *Sacrifice: Its Nature and Function,* trans. W. D. Halls. Chicago: University of Chicago Press. Orig. pub. 1898.

Hutchinson, Sharon. 1992. "The Cattle of Money and the Cattle of Girls among the Nuer, 1930–83." *American Ethnologist* 19(2):294–315.

———. 1996. *Nuer Dilemmas: Coping with Money, War, and the State.* Berkeley: University of California Press.

Johnson, Douglas H. 1994. *Nuer Prophets: A History of Prophecy from the Upper Nile in the Nineteenth and Twentieth Centuries.* Oxford, U.K.: Clarendon Press.

Johnson, Steven Lee. 1976. "An Analytical Approach to Rural Decision-Making: Interim Research Report." University of Nairobi, Institute of Development Studies, Working Paper 291.

———. 1979. "Micro-level Dependent Relations in Rural Kenya." Paper presented at the 55th Annual Meeting of the Central States Anthropological Society, Milwaukee, March 28–31, 1979.

———. 1980. *Production, Exchange, and Economic Development among the Luo-Abasuba of Southwestern Kenya.* Ph.D. thesis, Indiana University. Ann Arbor, Mich.: University Microfilms.

———. 1983. "Social Investment in a Developing Economy: Position-Holding in Western Kenya." *Human Organization* 42(4):340–46.

Johnston, Sir Harry. 1904. *The Uganda Protectorate,* vol. 2. London: Hutchinson and Co.

Jones, Karen. 1996. "Trust as an Affective Attitude." *Ethics* 107:4–25.

Josupeit, Helga. 2006. "Nile Perch Market Report" (August). Globefish, U.N. Food and Agricultural Organization, http://www.globefish.org/index.php?id=3073, on August 27, 2006.

Kenya. *Gazette.* Periodical. Nairobi.

Kenya, Central Bureau of Statistics. 1979, 1989, 1999. *Population Census.* Nairobi: Government Printer.

Kenya, Central Bureau of Statistics. *Integrated Rural Surveys.* Periodical. Nairobi.

Kenya, Colony and Protectorate of, *Annual Reports.* Nairobi: Government Printer.

Kenya, Ministry of Agriculture, Animal Husbandry, and Water Resources. 1956. *African Land Development in Kenya, 1946–55.* Nairobi: Government Printer.

Kenya, Ministry of Economic Planning; Central Bureau of Statistics. *Kenya Statistical Abstracts.* Annual. Nairobi: Government Printer.

Kenya, Ministry of Economic Planning and Institute of African Studies, University of Nairobi. 1986. *District Socio-Cultural Profiles Project: South Nyanza District.* Institute of African Studies, University of Nairobi.

Kenya, Ministry of Finance and Planning, Central Bureau of Statistics. 1977. *Rural Household Survey, Nyanza Province 1970–71*. Nairobi: Central Bureau of Statistics.

Kilbride, Philip, Collette Suda, and Enos Njeru. 2000. *Street Children in Kenya: Voices of Children in Search of a Childhood*. Westport, Conn.: Bergin and Garvey.

Kitching, Gavin. 1980. *Class and Economic Change in Kenya: The Making of an African Petite-Bourgeoisie*. New Haven, Conn.: Yale University Press.

Kokwaro, J. O. 1972. *Luo-English Botanical Dictionary*. Nairobi: East African Publishing House.

————. 1979. *Classification of East African Crops*. Nairobi: Kenya Literature Bureau.

Kongstad, Per, and Mette Monsted. 1980. *Family, Labour and Trade in Western Kenya*. Uppsala: Scandinavian Institute of African Studies.

Kopytoff, Igor. 1997. "Ancestors as Elders in Africa." In Roy Richard Grinker and Christopher B. Steiner, eds., *Perspectives on Africa*, pp. 412–21. Oxford, U.K., and Cambridge, Mass.: Blackwell. Orig. pub. 1971.

Kudhongania, A. W., D. L. Ocenodongo, and J. O. Okaronon. 1996. Anthropogenic Perturbations on the Lake Victoria Ecosystem. In Thomas C. Johnson and Eric O. Odada, eds., *The Limnology, Climatology and Paleoclimatology of the East African Lakes*, pp. 625–32. Amsterdam: Gordon and Breach.

Kuper, Adam. 1982. *Wives for Cattle: Bridewealth and Marriage in Southern Africa*. London and Boston: Routledge and Kegan Paul.

Leach, Edmund. 1982. *Social Anthropology*. Oxford and New York: Oxford University Press.

LeClair, Edward, and Harold Schneider, eds. 1968. *Economic Anthropology*. New York: Holt, Rinehart.

Lévi-Strauss, Claude. 1969. *The Elementary Structures of Kinship* (*Les Structures Elémentaires de la Parenté*). Rev. ed. Trans. James Harle Bell, John Richard von Sturmer, and Rodney Needham; ed. Rodney Needham. Boston: Beacon. Orig. pub. 1949.

LeVine, Robert A., and Barbara B. LeVine. 1966. *Nyansongo: A Gusii Community in Kenya*. New York: John Wiley.

LeVine, Robert A., S. Dixon, S. LeVine, A. Richman, P. H. Leiderman, C. H. Keefer, and T. B. Brazelton. 1996. *Child Care and Culture: Lessons from Africa*. Cambridge: Cambridge University Press. Orig. pub. 1994.

LeVine, Sarah. 1979. *Mothers and Wives: Gusii Women of East Africa*. Chicago: University of Chicago Press.

Leys, Colin. 1975. *Underdevelopment in Kenya*. London: Heinemann.

Lienhardt, Godfrey. 1961. *Divinity and Experience: The Religion of the Dinka*. Oxford: Clarendon Press.

————. 1975. "Getting Your Own Back: Some Themes in Nilotic Myth." In John Beattie and Godfrey Lienhardt, eds., *Studies in Social Anthropology*. Oxford: Clarendon Press.

Little, Peter D. 2003. *Somalia: Economy without State*. Oxford, Bloomington, and Hargeisa: James Currey, Indiana University Press, and Btec Books.

Little, Peter D., and Michael M. Horowitz. 1987. "Subsistence Crops *Are* Cash Crops: Some Comments with Reference to Eastern Africa." *Human Organization* 46(3):254–58.

Locke, John. 1952. *The Second Treatise of Government*. New York: Liberal Arts Press. Orig. pub. 1690.

Lockhart, Russell A., James Hillman, Arwind Vasavada, John Weir Perry, Joel Covitz, and Adolf Guggenbuehl-Craig. 1982. *Soul and Money*. Dallas, Tex.: Spring Publications.

Lohrentz, Kenneth P., and Alan C. Solomon. 1975. *A Guide to Nyanza Province Microfilm Collection, Kenya National Archives*. Part 3, section 10: Daily Correspondence and Reports, 1930–1963, vols. 1 and 2. Foreign and Comparative Studies, Eastern Africa Occasional Bibliography no. 26. Syracuse, N.Y.: Maxwell School of Citizenship and Public Affairs, Syracuse University.

Long, Norman. 1977. *Introduction to the Sociology of Rural Development*. London: Tavistock.

———, ed. 1985. *Family and Work in Rural Societies: Perspectives on Non-Wage Labour*. London: Methuen.

Lonsdale, John M. 1964. *A Political History of Nyanza, 1883–1945*. Ph.D. thesis, Cambridge University.

Lovejoy, Paul, and Toyin Falola, eds. 2003. *Pawnship, Slavery, and Colonialism in Africa*. Trenton, New Jersey: Africa World Press.

Luke, Nancy. 2006. "Exchange and Condom Use in Informal Sexual Relationships in Urban Kenya." *Economic Development and Cultural Change* 54(2):319–48.

Macgoye, Marjorie Oludhe. 1987. *Street Life*. Nairobi: Heinemann Kenya.

———. 2000. *Coming to Birth*. New York: Feminist Press at the City of New York. Orig. pub. 1986.

Maine, Henry. 1912. *Ancient Law*. London, Toronto: J. M. Dent and Sons. Orig. pub. 1861.

Malinowski, Bronislaw. 1922. *Argonauts of the Western Pacific*. New York: Dutton.

Mann, Charles K., Merilee S. Grindle, and Parker Shipton, eds. 1989. *Seeking Solutions: Framework and Cases for Small Enterprise Development Programs*. West Hartford, Conn.: Kumarian Press.

Marco Surveys Ltd. (for Ministry of Labour and Social Services). 1965. *A Baseline Survey of Factors Affecting Agricultural Development in Three Areas of Kenya (Samia, Kabondo, Bomet)*. Nairobi: Marco Surveys Ltd.

Marshall, Alfred. 1929. *Money, Credit and Commerce*. London: MacMillan. Orig. pub. 1923.

Marx, Karl. 1906. *Capital: A Critique of Political Economy*. Vol. 1: *The Process of Capitalist Production*. Chicago: Charles H. Kerr and Co. Orig. pub. 1867.

Mauss, Marcel. 1967. *The Gift*. Trans. Ian Cunnison. New York: Norton. Orig. pub. 1925.

Maxon, Robert M. 1989. *Conflict and Accommodation in Western Kenya: The Gusii and the British, 1907–1963*. Rutherford, Madison, and Teaneck: Fairleigh Dickinson University Press.

Mayer, Philip. 1949. "The Lineage Principle in Gusii Society." International African Institute Memorandum 24. London: Oxford University Press.

Mbogoh, S. G. (misattrib.). 1986a. "Land Tenure." In Kenya Ministry of Economic Planning and Development, and Institute of African Studies, University of Nairobi, *District Socio-Cultural Profiles Project: South Nyanza District*, pp. 38–50.

——— (misattrib.). 1986a. "Production Systems and Labour." In Kenya Ministry of Economic Planning and Development, and Institute of African Studies, University of Nairobi, *District Socio-Cultural Profiles Project: South Nyanza District*, pp. 28–37.

Mboya, Paul. 2001. *Luo—Kitgi gi Timbegi*. Trans. Jane Achieng. Nairobi: Atai Joint
Limited. [DhoLuo version: Kisumu: Anyange Press and East African Standard]. Orig.
pub. 1938.

Meillassoux, Claude. 1981. *Maidens, Meal and Money: Capitalism and the Domestic
Community*. Cambridge: Cambridge University Press. Orig. pub. 1975 as *Femmes, greniers
et capitaux*.

Meinhard, H. H. n.d. *Notes on the Bakuria*. Unpublished typescript.

Middleton, John. 1965. *Central Tribes of the North-East Bantu*. Ethnographic Survey of
Africa, East Central Africa, Part 5. London: International African Institute.

———. 1960. *Lugbara Religion: Ritual and Authority among an East African People*.
London: Oxford University Press, for International African Institute.

Migot-Adholla, Shem Edwin. 1977. *Migration and Rural Differentiation in Kenya*. Ph.D.
thesis, University of California at Los Angeles.

Moock, Joyce Lewinger. 1986. *Understanding Africa's Rural Households and Farming
Systems*. Boulder, Colo.: Westview.

Moore, Sally Falk. 1986. *Social Facts and Fabrications: "Customary" Law on Kilimanjaro,
1880–1980*. Cambridge: Cambridge University Press.

Morgan, W. B., and R. P. Moss. 1972. "Savannah and Forest in Western Nigeria." In R. M.
Prothero, ed., *People and Land in Africa South of the Sahara*, 20–27. London: Oxford
University Press.

Muga, Erasto. 1980. *Studies in Prostitution: East, West and South Africa, Zaire and Nevada*.
Nairobi: Kenya Literature Bureau.

Mutongi, Kenda. 1999. "Worries of the Heart: Widowed Mothers, Daughters and
Masculinities in Maragoli, Western Kenya, 1940–60." *Journal of African History*
40:67–86.

Mwangi, Meja. 1976. *Going Down River Road*. London: Heinemann.

Nakabayashi, Nobuhiro. 1981. "The Clan System and Social Change in Modern Isukha."
In Nobuhiro Nagashima, ed., *Themes in Socio-Cultural Ideas and Behavior among the Six
Ethnic Groups of Kenya*, 14–42. Tokyo: Hitotsubashi University.

Needham, Rodney. 1975. "Polythetic Classification." *Man* (new series) 10:349–67.

Nietzsche, Friedrich. 1969. *On the Genealogy of Morals* (Zur Geneologie der Moral). Trans.
Walter Kaufmann. Orig. pub. 1878.

———. *Ecce Homo*. 1969. Trans. Walter Kaufmann and R. J. Hollingdale. New York:
Vintage. Orig. pub. 1908.

North, Douglass C. 1981. *Structure and Change in Economic History*. New York: W. W.
Norton.

Northcote, G. A. S. 1907. "The Nilotic Kavirondo." *Journal of the Royal Anthropological
Institute* 38: 58–66.

———. 1909. "Letter and Annual Report to the Provincial Commissioner, Kisumu,
20 December." KNA/Syracuse University Microfilm 2801, Roll 37.

Nyangweso (pseud.). 1930. "Cult of Mumbo in Central and South Kavirondo." *Journal of
the East Africa and Uganda Natural History Society* 10(38):13–17.

Obama, Barack. 2004. *Dreams from My Father: A Story of Race and Inheritance*. New York:
Three Rivers Press. Orig. pub. 1995.

Obara, Dunstan A. 1979. "Cotton Production and Marketing in the Lake Victoria Basin of Kenya." Working paper 353 (rev.), Institute for Development Studies, University of Nairobi.

Ochieng', William Robert. 1975. *The First Word: Essays on Kenya History*. Nairobi: East African Literature Bureau.

Ocholla-Ayayo, A. B. C. 1976. *Traditional Ideology and Ethics among the Southern Luo*. Uppsala: Scandinavian Institute of African Studies.

———. 1979. "Marriage and Cattle Exchange among the Nilotic Luo." *Paideuma* 25:173–93.

———. 1980. *The Luo Culture: A Reconstruction of the Material Patterns of a Traditional African Society*. Wiesbaden: Franz Steiner Verlag.

Oculi, Okello. 1968. *Prostitute* (novel). Nairobi: East African Publishing House.

Odaga, Asenath Bole. 1971. *The Villager's Son*. Kisumu, Kenya: Lake Publishers and Enterprises.

———. 1980. *Thu Tinda! Stories from Kenya*. Nairobi: Uzima Press.

———. 2004. *Dholuo-English Dictionary*. Kisumu, Kenya: Lake Publishers and Enterprises.

Odede, Walter. 1942. "Luo Customs with Regard to Animals (with Particular Reference to Cattle)." *Journal of the East Africa and Uganda Natural History Society* 16:127–35.

Odinga, Oginga. 1967. *Not Yet Uhuru*. London: Heinemann.

Ogot, Bethwell A. 1967. *Peoples of East Africa: History of the Southern Luo*, vol. 1, *Migration and Settlement 1500–1900*. Nairobi: East African Publishing House.

———. 1971. "Reverend Alfayo Odongo Mango, 1870–1934." In Kenneth King and Ahmed Salim, eds., *Kenya Historical Biographies*, pp. 98–111. Nairobi: East African Publishing House.

———. 1999. *Reintroducing Man into the African World*. Kisumu, Kenya: Anyange Press.

Ogot, Grace. 1984. "Land Without Thunder." In Grace Ogot, *Land Without Thunder* (short stories), pp. 143–60. Nairobi: East African Publishing House. Orig. pub. 1968.

Ojwang, J. B., and J. N. R. Mugambi, eds. 1989. *The S. M. Otieno Case*. Nairobi: Kenya Literature Bureau.

Okoth-Ogendo, H. W. O. 1981. "Land Policy and Agricultural Production in Kenya: the Case of the Nyanza Sugar Belt." In H. W. O. Okoth-Ogendo, ed., *Approaches to Rural Transformation in Eastern Africa*, pp. 164–84. Nairobi: Bookwise.

Ominde, Simeon H. 1952. "The Luo Girl, from Infancy to Marriage." Nairobi: East African Literature Bureau. Also London: Macmillan.

———. 1963. *Land and Population in the Western Districts of Nyanza Province, Kenya*. Ph.D. thesis, University of London.

Onyango, Philista M., Katete Orwa, Aloys A. Ayako, J. B. Ojwang', and Priscilla W. Kariuki. 1991. *Research on Street Children in Kenya*. Report to Kenya Attorney General's Office.

Onyango-Ogutu, B., and A. A. Roscoe. 1981 [1974]. *Keep My Words: Luo Oral Literature*.

Ostrom, Elinor, and James Walker, eds. 2003. *Trust and Reciprocity: Interdisciplinary Lessons from Experimental Research*. New York: Russell Sage Foundation.

Oucho, John. 2002. *Undercurrents of Ethnic Conflict in Kenya.* Amsterdam: E. J. Brill.

Pala Okeyo, Achola. 1977. *Changes in Economy and Ideology: A Study of the Joluo of Kenya (With Special Reference to Women).* Ph.D. thesis, Harvard University.

Parkin, David J. 1969. *Neighbours and Nationals in an African City Ward.* Berkeley: University of California Press.

——. 1972. *Palms, Wine, and Witnesses: Public Spirit and Private Gain in an African Farming Community.* San Francisco: Chandler.

——. 1978. *The Cultural Definition of Political Response: Lineal Destiny among the Luo.* London: Academic Press.

——. 1980. "Kind Bridewealth and Hard Cash." In Comaroff, John, ed., *The Meaning of Marriage Payments.* New York: Academic Press, 197–220.

Parry, Jonathan, and Maurice Bloch, eds. 1989. *Money and the Morality of Exchange.* Cambridge: Cambridge University Press.

Paterson, Douglas B. 1984. *Kinship, Land and Community: The Moral Foundations of the Abaluhya of East Bunyore (Kenya).* Ph.D. thesis, University of Washington. Ann Arbor: University Microfilms International.

p'Bitek, Okot. 1971. *Religion of the Central Luo.* Nairobi, Kampala, Dar es Salaam: East African Literature Bureau.

——. 1985. "Acholi Proverbs." Nairobi: Heinemann Kenya.

Piot, Charles D. 1991. "Of Persons and Things: Some Reflections on African Spheres of Exchange." *Man* (new series) 26:405–424.

Polanyi, Karl. 1957. *The Great Transformation: The Political and Economic Origins of Our Time.* Boston: Beacon Press. Orig. pub. 1944.

Potash, Betty. 1978. "Some Aspects of Marital Stability in a Rural Luo Community." *Africa* 48(4):380–97.

——. 1986. "Widows of the Grave: Widows in a Rural Luo Community." In Betty Potash, ed., *Widows in African Societies: Choices and Constraints,* pp. 46–65. Stanford, Calif.: Stanford University Press.

Pottier, Johan P. 1986. "The Politics of Famine Prevention: Ecology, Regional Production and Food Complementarity in Western Rwanda." *African Affairs* 85(339):207–37.

Pratt, Beth Anne. 2003. *Childhood, Space and Children "Out of Place": Versions of Maasai Childhood in Monduli Juu, Tanzania.* Ph.D. thesis, Boston University.

Prince, Ruth J. 2004. "Shared Lives: Exploring Practices of Amity between Grandmothers and Grandchildren in Western Kenya." *Africa* (Journal of the International African Institute) 74(1): 95–120.

Rappaport, Roy A. 1999. *Ritual and Religion in the Making of Humanity.* Cambridge: Cambridge University Press.

Robertson, A. F. 1984. *People and the State: An Anthropology of Planned Development.* Cambridge: Cambridge University Press.

——. 1987. *The Dynamics of Productive Relationships: African Share Contracts in Comparative Perspective.* Cambridge: Cambridge University Press.

Rodney, Walter. *How Europe Underdeveloped Africa.* London and Dar es Salaam: Bogle L'Ouverture and Tanzanian Publishing House.

Rogerson, J. W. 1980. "Sacrifice in the Old Testament: Problems of Method and Approach." In M. F. C. Bourdillon and Meyer Fortes, eds., *Sacrifice*, pp. 45–60. New York and London: Academic Press.

Roscoe, John. 1915. *The Northern Bantu: An Account of Some Central African Tribes of the Uganda Protectorate*. London: Frank Cass.

Rotter, J. B. 1980. "Interpersonal Trust, Trustworthiness, and Gullibility." *American Psychologist* 35:1–7.

Rousseau, Jean-Jacques. 1967. "Discourse on the Origin and the Foundation of Inequality among Mankind" (*Discours sur l'origine et les fondements de l'inégalité parmi les hommes*). In J.-J. Rousseau, *The Social Contract and Discourse on the Origin of Inequality*, pp. 175–258. New York: Washington Square Press. Orig. pub. 1755.

Ruel, Malcolm J. 1965. "Religion and Society among the Kuria of East Africa." *Africa* 35:295–306.

Ruel, Malcolm. 1997. *Belief, Ritual and the Securing of Life: Reflexive Essays on a Bantu Religion*. Leiden, New York, and Koln: E. J. Brill.

Ruhlen, Merritt. 1991. *A Guide to the World's Languages*, vol. I: *Classification*. Stanford, Calif.: Stanford University Press.

Sagan, Eli. 1974. *Cannibalism: Human Aggression and Cultural Form*. New York: Harper and Row.

Sahlins, Marshall. 1974. *Stone Age Economics*. London: Tavistock. Orig. pub. 1972.

Sanderson, Stephen K. 1987. "Eclecticism and Its Alternatives." *Current Perspectives in Social Theory* 8:313–45. London: JAI Press.

Sangree, Walter. 1966. *Age, Prayer and Politics in Tiriki, Kenya*. London: Oxford University Press, for International African Institute.

Sauper, Hubert (author, dir.). 2004. *Darwin's Nightmare* (documentary motion picture). Paris: Mille et Une Productions.

Schwartz, Nancy. 1989. *World Without End: The Movements and Meanings in the History, Narratives, and "Tongue-Speech" of Legio Maria of African Church Mission among Luo of Kenya*. Ph.D. thesis, Princeton University.

———. 1995. "Contested Spaces, Contested Places: Anthropology's Necrology and the Rewriting of Lives in Western Kenya." Paper presented at the 38th Annual Meeting of the African Studies Association, Orlando, Florida, Nov. 3–6.

———. 2000. "Active Dead or Alive: Some Kenyan Views about the Agency of Luo and Luyia Women Pre- and Post-Mortem." *Journal of Religion in Africa* 30(4):433–67.

Schwartz, Ute, Aine Costigan, Elizabeth Ngugi, Joshua Kimani, Steven Moses, Frank Plummer. 1999–2000(?). "Knowledge, Attitudes, Practices and Prevalence of Sexually Transmitted Diseases among Female Sex Workers in Western Province, Kenya." Study for the Strengthening STD/HIV/AIDS Control in Kenya Project, University of Nairobi and University of Manitoba. Durham, N.C.: Family Health International.

Schwartz, Ute, Aine Costigan, Elizabeth N. Ngugi, Joshua Kimani, Steven Moses, J. Price, Frank Plummer. 2000. "Sexual Behaviour among Female Sex Workers and Male Company Workers in Western Province, Kenya." International Conference on AIDS, July 7–12; 14: abstract MoPeC6169. http://gateway.nlm.nih.gov/MeetingAbstracts/102251169.html, on September 3, 2006.

Scott, James C. 1976. *The Moral Economy of the Peasant*. New Haven, Conn.: Yale
University Press.

Seddon, D., ed. 1978. *Relations of Production: Marxist Approaches to Economic Anthropology*.
London: Frank Cass.

Shack, William, and Elliot P. Skinner, eds. 1979. *Strangers in African Societies*. Berkeley:
University of California Press.

Shadle, Brett L. 2000. "Girl Cases: Runaway Wives, Eloped Daughters and Abducted
Women in Gusiiland, Kenya, c. 1900–1965." Ph.D. dissertation, Northwestern
University.

———. 2002. "Patronage, Millennialism and the Serpent God Mumbo in South-west
Kenya, 1912–1934." *Africa* (Journal of the International African Institute) 72(1): 29–54.

———. 2003. "Bridewealth and Female Consent: Marriage Disputes in African Courts,
Gusiiland, Kenya." *Journal of African History* 44: 241–62.

Shakespeare, William. 1965. *Hamlet*, ed. Horace Howard Furness. New Variorum Edition.
New York: American Scholar Publications.

Sheridan, Michael. 2001. *Cooling the Land: The Political Ecology of North Pare, Tanzania*.
Ph.D. thesis, Boston University.

Shipton, Parker. 1984. "Strips and Patches: A Demographic Dimension in Some African
Landholding and Political Systems." *Man* (new series) 19:613–34.

———. 1988. "The Kenyan Land Tenure Reform: Misunderstandings in the Public
Creation of Private Property." In R. E. Downs and S. P. Reyna, eds., *Land and Society in
Contemporary Africa*, 91–135. Hanover, N.H.: University Press of New England.

———. 1989. *Bitter Money: Cultural Economy and Some African Meanings of Forbidden
Commodities*. American Ethnological Society Monograph 1. Washington, D.C.: American
Anthropological Association.

———. 1990a. "How Gambians Save—And What Their Strategies Imply for
International Aid." Policy, Research, and External Affairs WPS 395. Washington, D.C.:
Agriculture and Rural Development Department, The World Bank.

———. 1990b. "African Famines and Food Security: Anthropological Perspectives."
Annual Review of Anthropology 19:353–94.

———. 1991. "Time and Money in the Western Sahel: A Clash of Cultures in Gambian
Rural Finance." In Michael Roemer and Christine Jones, eds., *Markets in Developing
Countries: Parallel, Fragmented, and Black*, pp. 113–39, 235–44. San Francisco: ICS
Press, for the Harvard Institute for International Development and the International
Center for Economic Growth.

———. 1992a. "The Rope and the Box: Group Savings in Rural Gambia." In Dale W.
Adams and Delbert Fitchett, eds., *Informal Finance in Developing Countries*. Boulder,
Colo.: Westview Press.

———. 1992b. "Debts and Trespasses: Land, Mortgages, and the Ancestors in Western
Kenya." In M. Goheen and P. Shipton, guest eds., *Rights over Land: Categories and
Controversies*. Special issue of *Africa* (Journal of the International African Institute)
62(3):357–88.

———. 1994. "Land and Culture in Tropical Africa: Soils, Symbols, and the Metaphysics
of the Mundane." *Annual Review of Anthropology* 23:347–77.

———. 1995. "Luo Entrustment: Foreign Finance and the Soil of the Spirits in Kenya."
Africa 65(2): 165–96.

Shipton, Parker, and Miriam Goheen. 1992. "Understanding African Landholding: Power,
Wealth, and Meaning." Introduction to M. Goheen and P. Shipton, guest eds., *Rights
over Land: Categories and Controversies*. Special issue of *Africa* (Journal of the
International African Institute) 62(3):307–25.

Simmel, Georg. 1982. *The Philosophy of Money*. Trans. Tom Bottomore and David Frisby.
Boston, London, Melbourne, and Henley: Routledge & Kegan Paul. Orig. pub. 1978.

Southall, Aidan. 1952. "Lineage Formation among the Luo." *International African Institute
Memorandum* 26, pp. 1–43. London: Oxford University Press.

———. 1970. "Rank and Stratification among the Alur and Other Nilotic Peoples." In
Arthur Tuden and Leonard Plotnicov, eds., *Social Stratification in Africa*, pp. 31–46. New
York: Free Press.

Southall, Aidan, and P. C. W. Gutkind. 1980. "Marriage." In Erasto Muga, ed., *Studies in
Prostitution: East, West, and South Africa, Zaire and Nevada*, pp. 36–69. Nairobi: Kenya
Literature Bureau.

Spear, Thomas, and Richard Waller, eds. 1993. *Being Maasai: Ethnicity and Identity in East
Africa*. London: J. Currey; Athens: Ohio University Press.

Sperber, Dan. 1987. *On Anthropological Knowledge* (*le Savoir des anthropologues*).
Cambridge: Cambridge University Press. Orig. pub. 1982.

Stafford, R. L. 1967. *An Elementary Luo Grammar*. Nairobi: Oxford University Press.

Stam, N. 1910. "The Religious Conceptions of the Kavirondo." *Anthropos* 5(2):359–62.

Stambach, Amy. 2000. *Lessons from Mount Kilimanjaro: Schooling, Community, and Gender
in East Africa*. New York and London: Routledge.

Steele, Polly. 1999. *The Slate You Are: Emigré Disciples' Accounts of Islamic Instruction in
West Africa, and Its Bearing on Secular Schooling*. Ed.D. dissertation, Harvard University.

Stiansen, Endre, and Jane I. Guyer, eds. 1999a. *Credit, Currencies and Culture: African
Financial Institutions in Historical Perspective*. Uppsala, Sweden: Nordiska
Afrikainstitutet (The Nordic African Institute).

———. 1999b. "Introduction." In Stiansen, Endre, and Jane I. Guyer, eds., *Credit,
Currencies and Culture: African Financial Institutions in Historical Perspective*, pp. 1–14.
Uppsala, Sweden: Nordiska Afrikainstitutet (The Nordic African Institute).

Stichter, Sharon. 1982. *Migrant Labour in Kenya: Capitalism and African Response*. London:
Longman.

Swadener, Beth Blue, with Margaret Kabiru and Anne Njenga. 2000. *Does the Village Still
Raise the Child? A Collaborative Study of Changing Child-rearing and Early Education in
Kenya*. Albany: State University of New York Press.

Swindell, Ken. 1985. *Farm Labour*. African Society Today series, ed. Robin Cohen.
Cambridge: Cambridge University Press.

Swynnerton, R. J. M. 1955. "A Plan to Intensify the Development of African Agriculture
in Kenya." Nairobi: Department of Agriculture, Government Printer. Orig. pub. 1954.

Sykes, S. W. 1980. "Sacrifice in the New Testament and Christian Theology." In M. F. C.
Bourdillon and Meyer Fortes, eds., *Sacrifice*, pp. 61–83. New York and London:
Academic Press.

Sytek, William. 1972. *Luo of Kenya*. Hraflex Books Ethnocentrism Series. New Haven, Conn.: Human Relations Area Files.

Taylor, Christopher C. 1992. *Milk, Honey, and Money: Changing Concepts in Rwandan Healing*. Washington, D.C.: Smithsonian Institution Press.

Testart, Alain. 1998. "Uncertainties of the 'Obligation to Reciprocate': A Critique of Mauss." In Wendy James and N. J. Allen, eds., *Marcel Mauss: A Centenary Tribute*. Oxford and New York: Berghahn.

Thomas, Barbara P. 1985. *Politics, Participation and Poverty: Development through Self-Help in Kenya*. Boulder, Colo.: Westview Press.

———. 1988. "Household Strategies for Adaptation and Change: Participation in Kenyan Rural Women's Associations." *Africa* 58(4):401–22.

Thomas, Lynn M. 2003. *Politics of the Womb: Women, Reproduction, and the State in Kenya*. Berkeley and London: University of California Press.

Twongo, T. 1996. "Growing Impact of Water Hyacinth on Nearshore Environment on Lakes Victoria and Kyoga (East Africa)." In Thomas C. Johnson and Eric O. Odada, eds., *The Limnology, Climatology and Paleoclimatology of the East African Lakes*, pp. 633–42. Amsterdam: Gordon and Breach.

United Nations International Labour Office (UNILO). 1972. *Employment, Incomes and Equality: A Strategy for Increasing Productive Employment in Kenya*. Geneva: International Labour Office.

Vail, Leroy. 1989. *The Creation of Tribalism in Southern Africa*. Berkeley and Los Angeles: University of California Press.

Von Mises, Ludwig. 1981. *The Theory of Money and Credit*. Trans. H. E. Batson. Indianapolis: Liberty Classics. Orig. pub. 1934.

Von Pischke, J.D., Dale W. Adams, and Gordon Donald, eds. 1983. *Rural Financial Markets in Developing Countries: Their Use and Abuse*. Baltimore, Md., and London: Johns Hopkins University Press, for the World Bank.

Wagner, Gunter. 1949. *The Bantu of North Kavirondo*. Vol. 1. London: Oxford University Press.

———. 1956. *The Bantu of North Kavirondo*. Vol. 2. London: Oxford University Press.

Waligorski, Andrzej. 1952. "Rozpoynanie i Wybor Gleb Pod Uprawe Warod Nilockich Luo" (The Choice of Soil for Cultivated Lands among the Nilotic Luo). Wrocklaw: *Lud*. (39): 181–227. Summarized in *African Abstracts* 16, no. 4 (1965): 182.

———. 1970. "Les Marchés des Luo vers 1946–1948." *Africana Bulletin* (Warsaw) 11:9–24.

Waller, Richard D. 1988. "*Emutai*: Crisis and Response in Maasailand 1883–1902." In Douglas Johnson and David Anderson, eds., *The Ecology of Survival: Case Studies from Northeast African History*, pp. 73–114. Boulder, Colo.: Westview; London: Lester Crook.

Wallman, Sandra. 1996. *Kampala Women Getting By: Wellbeing in the Time of AIDS*. London, Kampala, and Athens: James Currey, Fountain Publishers, and Ohio University Press.

Weiner, Annette. 1992. *Inalienable Possessions: The Paradox of Keeping-While-Giving*. Berkeley: University of California Press.

Weisner, Thomas S. 1997. "Support for Children and the African Family Crisis." In

Thomas S. Weisner, Candice Bradley, and Philip L. Kilbride, eds., *African Families and the Crisis of Social Change*, pp. 20–44. Westport, Conn., and London: Bergin and Garvey.

Weisner, Thomas S., Candice Bradley, and Philip L. Kilbride, eds. 1997. *African Families and the Crisis of Social Change*. Westport, Conn., and London: Bergin and Garvey.

Whisson, Michael G. 1961. "The Rise of Asembo and the Curse of Kakia." East African Institute of Social Research *Conference Papers*. Makerere University College, Kampala.

———. 1962a. "The Will of God and the Wiles of Men: an Examination of the Beliefs Concerning the Supernatural Held by the Luo with Particular Reference to their Functions in the Field of Social Control." East African Institute of Social Research *Conference Papers*, Makerere University College, Kampala.

———. 1962b. "The Journeys of the JoRamogi." East African Institute of Social Research *Conference Papers*, Makerere University College, Kampala.

———. 1964. *Change and Challenge: A Study of the Social and Economic Changes among the Kenya Luo*. Nairobi: Christian Council of Kenya.

White, Luise. 1990. *The Comforts of Home: Prostitution in Colonial Nairobi*. Chicago: University of Chicago Press.

Whyte, Susan Reynolds. 1997. *Questioning Misfortune: The Pragmatics of Uncertainty in Uganda*. Cambridge: Cambridge University Press.

———. 1990. "The Widow's Dream: Sex and Death in Western Kenya." In Michael Jackson and Ivan Karp, eds., *Personhood and Agency: The Experience of Self and Other in African Cultures*. Uppsala Studies in Cultural Anthropology 14. Uppsala: Uppsala University.

Wilson, Gordon M. 1967. "Homicide and Suicide among the Joluo of Kenya." In Paul Bohannan, ed., *African Homicide and Suicide*. New York: Atheneum. Orig. pub. 1960.

———. 1968. *Luo Customary Law and Marriage Laws Customs* [sic]. Nairobi: Government Printer. Orig. pub. 1961.

———. (n.d.) *Rural Development Survey in Three Areas of Kenya: An Evaluation of Three Years of Rural Development and Change at Samia-Kabondo-Bomet Experimental Pilot Projects, 1965 compared with 1968*. Report to Kenya Ministry of Cooperatives and Social Services, Department of Community Development, Nairobi. Institute of Development Studies Library, University of Nairobi.

Wipper, Audrey. 1977. *Rural Rebels: A Study of Two Protest Movements in Kenya*. Oxford and Nairobi: Oxford University Press.

———. 1984. "Women's Voluntary Associations." In Margaret Jean Hay and Sharon Stichter, eds. *African Women South of the Sahara*, pp. 59–86. Burnt Mill, Essex, England: Longman.

Wittgenstein, Ludwig. 1958. *Preliminary Studies for the "Philosophical Investigations," Generally Known as the Blue and Brown Books*. Oxford: Blackwell.

———. 1984. *Culture and Value* (*Vermischte Bemerkungen*). Trans. Peter Winch. Chicago: University of Chicago Press. Orig. pub. 1977.

World Bank. 2006. Poverty Map, Nyanza Province, Kenya. http://www.worldbank.org/research/povertymaps/kenya/ch4.9.pdf, May 31, 2006.

Zelizer, Viviana. 1994. *The Social Meanings of Money*. New York: Basic Books.

Index

abduction of brides, Luo patterns of, 138–39, 240n44

Abila, Richard, 60

Abrahams, R. G., 170

Abwunza, Judith, 148, 158, 236n21, 241

accounts and accounting: debt and credit in terms of, 9–10; in marriage debts, 134, 239n32

Acholi culture and society, 98

age: differences in marriage, 130–32; divorce patterns and, 137–39; hierarchy, 74; importance of, 49; principles of grading and initiation, 49, 72. *See also* government and governance; hierarchy; sequencing

agriculture: labor exchanges and, 104–7; Luo knowledge of, 44–45, 53–56, 97–98, 226n2; money lending and, 114–16. *See also* food production

ancestors: cultural role of, 158–59, 220, 241n1; and inheritance, 183; and lake fishing, 58; and land, 49–53; and sacrifice, 58, 187–96, 216–17, 203–4. *See also* patriliny

animal exchanges (and slaughter): in bridewealth payments, 122–26, 131–32; divorce and return of, 136–39, 240nn38–41; earmarking of bridewealth payments and, 132–34, 239nn22–24; entrustment practices and, 91–94; equivalencies in, 94–97; fosterage practices and, 83–87, 230n2; at funerals, 160–72, 243n7; as homicide compensation, 168–70; human pledging and, 139–41; inheritance decisions and, 176–86, 243nn3–

4; inheritance of debts involving, 164–72; interlinking marriage patterns and, 126–28; labor exchanges and, 104–7, 234n9, 235n11; in Luo culture, 55–56, 154–57, 228nn21, 27; marriage rituals involving, 129–32; offspring exchanges, 94–97, 233n26; sacrificial rituals involving, 189–204, 245n14; sex separation in, 94–97; symbolic meanings of animals and, 237n7, 246n3

Ardener, Shirley, 38

association: trust and, 34–37. *See also* friendship; group membership; identity; labor groups; politics

asymmetry, in entrustment, 207–8, 221

Atieno Odhiambo, E. S., 40, 86, 150–51, 202

ayie (cash payment), 131–32

Baier, Annette, 34

Banholzer, P., 189

baraza (political assembly), 73

bargaining, in Luo culture, 64–66

Barth, Fredrik, 24

Bateson, Patrick, 218, 246n2

Benedict, Ruth, 21–23

betrayal: commercial sex and, 141–49; trust and risk of, 34–37

birds, entrustment relations in, 246n2

"bitter money": exclusion of, in bridewealth payments, 124; and roosters, 234; sacrificial ritual and, 195–96; separation of, 124, 212; shared responsibility for, 171–72